Common Sense and Science
from Aristotle to Reid

T0178799

Common Sense and Science from Aristotle to Reid

Benjamin W. Redekop

ANTHEM PRESS

Anthem Press
An imprint of Wimbledon Publishing Company
www.anthempress.com

This edition first published in UK and USA 2021
by ANTHEM PRESS
75–76 Blackfriars Road, London SE1 8HA, UK
or PO Box 9779, London SW19 7ZG, UK
and
244 Madison Ave #116, New York, NY 10016, USA

First published in the UK and USA by Anthem Press 2020

British Library Cataloguing-in-Publication Data
A catalogue record for this book is available from the British Library.

Library of Congress Control Number: 2020952974

ISBN-13: 978-1-78527-980-5 (Pbk)
ISBN-10: 1-78527-980-7 (Pbk)

This title is also available as an e-book.

CONTENTS

ACKNOWLEDGMENTS

This book has benefited from the insights and encouragement of many people and the support of a variety of institutions. I wish to thank former president Gordon Johnson and the fellows of Wolfson College, Cambridge, for support at the start of this project when I was a research fellow of the College. The late Istvan Hont was an important intellectual mentor during my time at Cambridge and I am most grateful that he took me under his wing during those years. The Cambridge University Library staff was very helpful, as were the staff of King's College, Aberdeen; the University of British Columbia; Oxford University; and the University of Michigan. Gordon Graham, Paul Gorner, and Maria-Rosa Antognazza provided help and guidance as this project was first germinating. I would also like to thank John Wright, Terence Cuneo, René van Woudenberg, Richard Little, and Rebecca Copenhaver for feedback on segments of this work. Paul Wood and Knud Haakonssen provided help, support, and guidance along the way, and their intellectual contributions to the topics covered in this book will be evident to readers who know their work. I would also like to thank the anonymous reviewers who took the time to review and make thoughtful comments and suggestions on the proposal and draft manuscript.

At Christopher Newport University I have received helpful feedback and experienced friendly collegiality from Brent Cusher, Nathan Harter, Bob Colvin, Quentin Kidd, Ed Brash, Roberto Flores, and Jon White, among others. Harvey Mitchell, Stephen Straker, and Allan Smith at the University of British Columbia provided important intellectual guidance and support during my years there. I have benefited from many conversations about cognitive science and cognitive psychology with my good friend Jim Enns, often in the midst of rock-climbing expeditions. I owe a debt of gratitude to Steven Spalding, who provided research and translation assistance with some of the French language sources used in this book. Thanks to Michael Callahan for sharing his friendship and love of history over many years. Ben Lynerd provided very helpful advice and encouragement as this project was nearing its end, and longtime friend Doug Balzer kept me going with regular reminders that I needed to finish the dang thing so he could read it. I would like to thank my wife Fran and daughter Katarina for their love and support during the two decades in which this book took shape. Finally, Ed Hundert deserves my highest gratitude for his long-standing support and mentorship. His stamp on my thinking and approach to intellectual history is doubtless evident on every page of this book, and it is to him that this book is dedicated.

Research for this book was supported by grants and fellowships from the Social Sciences and Humanities Research Council of Canada, Kettering University, and Wolfson College, Cambridge. Research and writing was made possible by sabbaticals funded by Kettering University and Christopher Newport University. Parts of Chapter 2, Chapter 8, and the Epilogue were originally published in the following works, reprinted here with permission: Benjamin W. Redekop, "Thomas Reid and the Problem of Induction: From Common Experience to Common Sense," *Studies in History and Philosophy of Science* 33A.1 (2002): 35–57; Benjamin W. Redekop, "Common Sense and Science: Reid Then and Now," *Reid Studies* 3.1 (1999): 31–47; Benjamin W. Redekop, "Reid's Influence in Britain, Germany, France, and America," in Terence Cuneo and René van Woudenberg (Eds.), *The Cambridge Companion to Thomas Reid* (Cambridge: Cambridge University Press, 2004), 313–39.

INTRODUCTION

Thinkers have been pondering the nature of common sense, and its relationship to science and scientific thinking, for a very long time. In the ancient world, "scientific" knowledge (*epistêmê* in Greek, *scientia* in Latin) emerged as a counterpoint to everyday understanding and common opinion, until Aristotle produced a reconciliation of the two that set the course for scientific thought for the next two millennia. It was not until the early-modern period, when the New Science of Copernicus, Galileo, and Newton emerged triumphant, that common sense and its relationship to science again became problematic, remaining so to this day. This book is about this fraught relationship and about the early-modern thinkers who sought to address it, culminating in the thought of the philosopher Thomas Reid (1710–1796), the preeminent figure in the Scottish school of common-sense philosophy. It is a story full of fascinating twists and turns but is ultimately about the perennial quest to understand how the human mind is able to gain credible and reliable knowledge about the self, nature, other human beings, and God.

It is my contention that if we can understand the historical interplay between common sense and science, and the emergence of "common sense" as a contested term in scientific and philosophical discourse, we will have a better grasp of some of the fundamental and enduring problems besetting the relationship between science and society, that is, the problem of the public understanding of science. In some parts of the world, including particularly the United States, questions that have been long settled in the scientific community—for example, global warming, evolution by natural selection, the value of vaccines—remain controversial in the larger public arena. A disconcerting gap persists between everyday knowledge and understandings, and well-established scientific theories and facts. While there are many factors—economic, religious, political—contributing to this state of affairs, it builds upon a mismatch between our everyday, commonsensical judgments and intuitions, and the discoveries and methods of modern science. This book helps readers to better understand the fundamental contours of this relationship and why common sense and science may not be at odds after all.

The modern philosophical conception of common sense arose in response to the skepticism and materialism unleashed by thinkers steeped in the methods and perspectives of the New Science, as part of a broad effort to connect higher thought with everyday perceptions and processes of the human mind, and by extension to the rising "commons" of Europe and America. This dynamic relationship between common sense and the rise of modern science has gone largely unrecognized in the history of ideas, and this study aims to bring it to light, while also providing an overview of the common-sense philosophical tradition—in all its various and sundry forms—that stretches all the way from Aristotle to the present day.

"Common sense," as a term of both popular and elite discourse, therefore has a history, and that history is related, at least in part, to modern science and the philosophical systems that arose along with it. This book tells that story, and it does so by taking seriously thinkers who have often received short shrift in the history of ideas. Until quite recently, many of the thinkers covered in this book have garnered little attention from scholars, despite the fact that some of the principal ideas and perspectives of this intellectual tradition have been validated by modern scientific research (discussed further in the Epilogue). Towering figures like Plato, Hobbes, Locke, and Hume thus play a supporting role in this story, having to make way for neglected yet notable thinkers such as Herbert of Cherbury, Henry More, Robert Ferguson, Henry Lee, Claude Buffier, the Third Earl of Shaftesbury, Francis Hutcheson, George Turnbull, Lord Kames, and Thomas Reid, among many others.

These latter thinkers form part of a long tradition of reflection on the important question of common sense and its relationship to science and higher thought, culminating—rather than beginning, as is often supposed—in the work of Reid. As we shall see, Reid's *An Inquiry into the Human Mind, on the Principles of Common Sense*, first published in 1764, was actually a capstone on more than a century of thinking about common sense and its relation to scientific thinking. That is one reason why this book culminates, rather than begins, with Reid. While it is surely overstating the case to suggest that most of the major philosophical issues regarding the nature of common sense and its relation to modern science were worked through by Reid's death in 1796, they were at least on the table and in many cases fully articulated. Thus if we understand the story from Aristotle to Reid, we are well positioned to make sense of the basic contours of the modern relationship between common sense and science, along with the various permutations of "common sense" as a term of modern social and political discourse.[1]

What is meant by the protean term "common sense" will emerge in the course of this study in relation to the specific contexts in which it was used.

Nevertheless, there are some basic definitions and common features of the term that can be stated at the outset. "Common-sense *experience*" aims to capture the grounded nature of Aristotle's epistemology, but I will also use the term whenever I want to emphasize the experiential elements of common sense. I will also use the terms "common sense," "common experience," "common knowledge," "everyday experience," "common opinion," and "intuition," among other terms, as required by the context. "Common sense" will ordinarily refer to the commonly held, seemingly self-evident perceptions and judgments of the average person, and hence of the local and/or larger human community. As such, the term implies the existence of a *faculty* of common sense that is part of the mental "furniture" of all healthy human beings, to borrow an expression used by George Turnbull (Chapter 7). Since the term carries with it both communal ("common") and perceptual ("sense") connotations and since it often carries a strong implication of intuitive or self-evident knowledge, its meaning will necessarily be somewhat elastic, depending on context.

Nicholas Rescher offers three definitions of "common sense" as a term of philosophy. *Observational* common sense refers to the classical sensus communis, involving "the collection and coordination of the deliverances of the external senses with special reference to those features of things that are accessible to more than one of our senses." *Judgmental* common sense is invoked in regard to "matters that are obvious and evident to anyone of sound understanding [...] on the basis of everyday experience, without elaborate reasoning, calculation, or investigation." It is a form of judgment. *Consensual* common sense involves matters of fact that "everyone knows and with respect to which there is a universal (or near-universal) agreement of people's opinion."[2] This would be what is generally understood as "common knowledge" and will also appear in the present study as "common notions," "common consent," or "universal consent," depending on the context and with some variation in meaning.

Although this taxonomy is not exhaustive, and (as Rescher notes) there is some conceptual overlap between these terms, it provides us with a sense of the ways in which contemporary philosophers understand the term and how it has been used. All three forms of common sense outlined by Rescher will be covered in this study. For his part, Reid spoke of "principles" of common sense that are specific, innate, automatic judgments of our minds made in the course of everyday experience and as such help us to make sense of the world, both physical and social (or moral)—the principle of causality being the most pertinent to the present study. As will become clear, although Reid presented a fully developed and articulated understanding of a faculty and principles of common sense, a variety of previous thinkers had grappled, in a similar fashion, with the notion that we come equipped with innate mental tools (nowadays we might say mental "modules" or "hardware") to make sense of the world, whether or not they used the term "common sense."

* * *

Before plunging into this story, however, it will be useful to briefly review contemporary thinking on the relationship between common sense and science, a topic to which we will return in the Epilogue. Generally speaking, there are two schools of thought on this relationship. One emphasizes the differences between common sense and science, and the strangeness of modern science to the untutored intellect. In this view, modern science arose with the ancient Greeks and goes against the grain of our normal, intuitive, egoistic, or "natural" ways of thinking. Common sense evolved to help us survive and navigate everyday circumstances, not to know the world scientifically, and is in fact a quite different form of knowledge. Proponents of this position tend to maintain that common sense is relative to culture and upbringing: that there are no universal, innate aspects of common sense, or if there are, they are irrelevant to scientific thinking.

The contrary position acknowledges differences between common sense and science but seeks to demonstrate the existence of some form of continuity between the two. Some thinkers maintain that common sense evolves and adjusts itself to science, while others maintain that there are universally prevalent, innate perceptions (or intuitions) that form a substructure for common sense and all higher forms of reasoning, including science. There is thus a continuum stretching between the intuitions that result from the way we are wired to those that come as a result of experience and shared cultural understandings. In this view, it is wrong to suppose that there is a radical break between common sense and scientific thinking. Without basic common-sense perceptions, we would never be able to make sense of the world at all, in either everyday *or* scientific terms.

However, even those that emphasize the gulf between common sense and science often recognize that there is some overlap between the two, while those that stress their continuity allow that scientific thinking goes far beyond common-sense truths and intuitions and can often violate them, at least initially. Most everyone accepts some form of complementarity between the two, that is, that they are each valid forms of knowing in their own sphere: common sense is crucial to our everyday functioning and survival in physical and social worlds, while scientific thinking digs below the surface of our intuitions and commonly accepted "truths" to reveal underlying causal processes that may go against our intuitive or culturally conditioned understandings.

An example of a work that emphasizes the differences between common sense and science is Lewis Wolpert's *The Unnatural Nature of Science* (1993). According to Wolpert, "The world is not constructed on a common-sensical basis [...] Scientific ideas are, with rare exceptions, counter-intuitive: they cannot be acquired by simple inspection of phenomena and are often outside of everyday experience."[3] Science is a rare bloom that arose only once, in ancient Greece, and is exemplified by the works of Euclid and Archimedes. Technology, which is more directly related to common sense and is aimed at our survival, evolved separately from science through the nineteenth century. Ironically, "[Wolpert's] own position, philosophically, is that of a common-sense realist. I believe that there is an external world which I share with others and which can be studied."[4] There is thus an inherent ambiguity and paradox in Wolpert's central thesis. This paradox—the scientist who distrusts common sense yet relies upon it to make sense of the world—is a central subject of this book. As we will see, the tradition of thought analyzed in the following chapters grappled with this and other related problems in an illuminating and productive fashion and, preeminently in the work of Reid, provided a coherent account of common sense and its relationship to science that is in accord with the findings and perspectives of modern cognitive science.

Published in the same year as Wolpert's book, Alan Cromer's *Uncommon Sense: The Heretical Nature of Science* advances an argument that is very similar to Wolpert's. Cromer uses the work of Piaget to provide a psychological explanation for why scientific thinking goes against common sense: human beings are fundamentally egocentric and hence have a hard time forming a truly objective, impersonal picture of the world. According to Cromer, the egocentric view is the common-sense view, while the objective view is the view of science. Scientific thinking is "analytic and objective, [and] goes against the grain of traditional human thinking, which is associative and subjective."[5] Cromer's focus on Piaget supports his anti-innatist stance on human mental development. As such, he ignores the mounting evidence that Piaget was wrong: humans do in fact come into the world equipped with mental structures and intuitions that enable scientific thinking and the transcendence of egocentrism.[6] It is a rather one-sided argument, and the book suffers for it.

A recent contribution to this school of thought is Duncan Watts' *Everything Is Obvious, Once You Know the Answer. How Common Sense Fails Us* (2011). The main thrust of this book and books like it (e.g., works by Nassim Nicholas Taleb[7]) is that our common-sense beliefs and intuitions are fundamentally misleading in a complex world. It is a genre of skepticism, aimed at helping us recognize the limitations of our mental abilities, particularly when it comes to large and complex social, economic, and physical systems. Such works are useful in that they can help us avoid mistakes in our thinking. The basic message is that we come equipped to make sense of the world in very practical, everyday terms, aimed at social and physical survival. We are not so good at the very large and the very small and at figuring out cause and effect relationships beyond simple observable events.

This line of argumentation supports the idea that there is a fundamental gap between common sense and modern science. The narrative structure of the book leads Watts to elide or ignore the basic universal features of the mind that make scientific reasoning possible in the first place, instead focusing on common sense as a cultural construct. If the subtitle of your book is "How common sense fails us," then there is little incentive to provide a nuanced picture of common sense.

* * *

A more subtle discussion of the relationship between common sense and science is found in Nicholas Rescher's *Common Sense: A New Look at an Old Philosophical Tradition* (2005; cited above). For Rescher, common sense is a kind of "cognitive minimum" preeminently having to do with meeting human needs. The "Achilles' heel" of common sense "is not its rightness but its limited range."[8] Common-sense statements are more secure but less defined and detailed than scientific ones. Textbook science bridges the gap between common sense and science with what Rescher calls "a realism of the middle range": it is a kind of "halfway house" between common sense and science, not the self-evident generalizations of common sense but closer and more allied to it than cutting-edge science.[9] The overall implication is that there is continuity between the two, rather than a gap. Rescher goes on to make the Reidian argument that common sense gives us facts about the world that philosophy violates at its own expense, and he concludes that scientific knowledge complements and supplements common-sense knowledge rather than abolishing or replacing it.[10]

In a similar vein, John Ziman argues at length in *Real Science* (2000) for the continuity and connections between common sense and science, even if

> science is not just a systematic enlargement of "folk science" or "common sense writ large." On the contrary, it is in its conformity to the *small print* of common sense that science is distinctive […] Taken one by one, the cognitive norms [of science] that have to be satisfied—accuracy, specificity, reproducibility, generality, coherence, consistency, rigor, and so on—are all perfectly commonsensical: but they are seldom applied simultaneously outside science.[11]

The theories that emerge can indeed conflict with existing common-sense beliefs about the life world, but over time become part of what "everybody knows." Scientific knowledge hence ultimately returns to its roots in everyday knowledge. "Even its grandest theoretical paradigms are inferred and rooted in down-to-earth empirical 'facts' and always have to incorporate uncritically a great deal of 'taken for granted' life-world knowledge."[12]

In an essay published in 1979, Michael Dummett similarly argues that there is no sharp dividing line between common sense and science, since in his view common sense evolves along with science. He doesn't think there is an unchanging common sense "theory of the world" not affected by science. Modern science and technology inevitably seep into everyday consciousness. Physical theory has "grown from the effort to formulate laws governing everyday phenomena, describable in terms of observable qualities: there is a continuous development from the steps we are forced to take even in everyday life to frame an adequate description of the world we observe and live in to the abstruse physics of today."[13]

This is a fairly strong statement, given that one of the most "abstruse" branches of modern physics, quantum mechanics, has posed a formidable challenge to the idea that there is a connection between common sense and science. The most commonly cited issue, as stated by Nick Huggett, is the following:

> Even a single quantum particle will typically fail to have a definite location or a sharp trajectory. If the momentum of a particle is sufficiently well-defined then by the Heisenberg uncertainty relations, it won't even be approximately localized. If there are several particles, then things are even worse: their wave functions may overlap, and we will be unable to distinguish particles by their distinct locations. But our common-sense notions seem to rely on continuously distinct trajectories for differentiating objects.[14]

Thus, quantum mechanics affronts our common-sense notion that individual entities retain a unique and locatable identity.

However, Huggett explores interpretations of quantum mechanics that can be reconciled to common-sense understandings of identity, concluding that "Bohmian mechanics shows very clearly that the empirical results of [quantum mechanics] do not, by themselves, entail a fundamental shift in the metaphysics of individuality." Bohmian particles have continuous "classical" trajectories and do not all have the same state and are at different places, features that are reconcilable with common sense. "What this shows is that the striking empirical features of the quantum world do not demand a completely new metaphysical outlook with regard to identity: if common sense is to be assaulted it must be done from a particular (albeit standard) understanding of the theory."[15] In other words, some interpretations of quantum mechanics would seem to accord with common sense better than others. In a similar vein, Peter Forrest suggests that quantum mechanics can be reconciled with common sense using a "Wigner-Dirac" interpretation.[16] Nicholas Maxwell for his part argued in 1966 that micro-level "physicalist" explanations that may violate common sense are no more or less "real" than macro-level common-sense understandings of the world, since they are in fact different *kinds* of explanation; common-sense understandings are "real" and objective to the degree that they are shared by human beings with their own unique perceptual equipment, while physicalist explanations are real to the extent that they are intelligible to all *rational* beings, which can include things like computers.[17]

Such works take their place in a literature, reaching all the way back to Bertrand Russell's *ABC of Relativity* (first published in 1925 and still in print),[18] which has sought to reconcile the often bewildering discoveries of modern physics with common-sense understandings of the world. This literature reached a peak of sorts during the 1950s with *Science and Common Sense* (1951) by James Conant and *Science and the Common Understanding* (1953) by J. Robert Oppenheimer, books still worth reading today.[19] We will return to consider these and other such works in greater detail in the Epilogue.

A common approach to understanding the difference between common sense and science is to differentiate their object. Kathleen Wilkes, for example, suggests that the difference between common sense and scientific psychology is that the former searches for particular, context-specific, prudential explanations while the latter seeks systematic, general (or universal) theories.[20] Rescher makes a similar point: "In everyday-life communication, where we are deeply concerned to protect our credibility and trustworthiness, we value security over informativeness." As such, it is perfectly acceptable to be less than precise in everyday situations; it matters not the exact size or nature of the rock falling down the cliff toward us, only that a rock is coming at us, and this fact needs to be communicated quickly. "In science, on the other hand, we value generality and precision over security."[21] That is to say, when "doing science" we are interested in making very precise, falsifiable statements about the nature of the rock, or perhaps the fossil embedded in the rock. This difference helps to explain why scientists often have trouble communicating scientific data to the public at large and why scientific knowledge can be twisted to suit nonscientific agendas: scientific knowledge favors nuance and resists overgeneralization and as such is easily reconfigured to suit social, political, and other practical, nonscientific ends.

Taking a somewhat different tack on the relationship between common sense and science, Nobel Prize-winning psychologist Daniel Kahneman explores what psychologists call "dual-processing" theories and what he calls "System 1" and "System 2," the former being our everyday, automatic, "fast" intuitive mental abilities, the latter our more conscious, effortful, "slow" rational capabilities. This differentiation between commonsensical and scientific forms of thinking is more a function of timing and difficulty than of object per se. As Kahneman describes it, when we are confronted with a new situation or problem, the "intuitive machinery" of System 1 does the best it can to find a solution, based on a variety of intuitive understandings (whether acquired or innate) and automatic heuristic tools. If an adequate solution is not found, "we often find ourselves switching to a slower, more deliberate and effortful form of thinking"—System 2. "In the picture that emerges from recent research, the intuitive System 1 is more influential than your experience tells you, and it is the secret author of many of the choices and judgments that you make."[22] System 1

> is generally very good at what it does: its models of familiar situations are accurate, its short-term predictions usually accurate as well, and its initial reactions to challenges are swift and generally appropriate. System 1 has biases, however, systematic errors that it is prone to make in specified circumstances [...] it sometimes answers easier questions than the one that is asked, and it has little understanding of logic and statistics.[23]

Common sense is thus conceived as our normal way of making sense of the world, adequate to many of its challenges yet riddled with biases and shortcuts that may be misleading. Our more conscious, effortful System 2 can correct the errors of System 1, but it is lazy and thinks itself less dependent on System 1 than it truly is.[24] Thus although Kahneman dwells on the ways in which we can be misled by common sense, he underlines its importance and centrality in our thought processes and presents it as a platform—a little rickety at times, to be sure—for higher, more rational, and deliberate forms of thinking. System 1 effortlessly originates "impressions and feelings that are the main sources of the explicit beliefs and deliberate choices of System 2," including basic intuitive perceptions of physical objects and the agency of humans, as well as learned associations between ideas. "In summary, most of what you (or your System 2) think and do originates in your System 1, but System 2 takes over when things get difficult, and it normally has the last word."[25]

* * *

We will return to these issues in the Epilogue. For now it is sufficient to note that the ideas and perspectives advanced by contemporary thinkers on the relationship between common sense and science are rooted in the history we are about to embark upon; and while providing a point of *entrée* into this history, they are also illuminated by it. Our story begins in Chapter 1 with common sense in Plato, Aristotle, and subsequent scientific thinking before Copernicus. Chapter 2 examines how the early-modern rejection of Aristotelianism and the rise of the New Science of Bacon, Kepler, Galileo, Boyle, and Newton displaced scientific thinking from its roots in common-sense experience. Chapter 3 examines early forms of intuitionism—including the thought of Herbert of Cherbury and René Descartes—that emerged in response to the skepticism and materialism that arose in tandem with the New Science. Chapter 4 carries the story forward through analysis of innatist responses to the thought of Thomas Hobbes and John Locke. Chapter 5 steps back to look at eighteenth-century employments of the term "common sense" by a variety of writers and thinkers, while Chapter 6 examines "common sense" and "moral sense" in the works of Claude Buffier, Frances Hutcheson, and Joseph Butler, among others. Chapter 7 focuses on common sense and the "science of man" in the writings of George Turnbull and Henry Home (Lord Kames), Thomas Reid's teacher and friend, respectively. Chapter 8 provides a contextual analysis of the thought of Reid, who, as an epistemologist and phi-losopher of science concerned about the modern rupture between common sense and science, provided a reconceptualization of Aristotelian "common experience" in his philosophy of common sense.

For its part, the Epilogue considers Reid's subsequent influence in Britain, Germany, France, and America, before placing his thought into the context of contemporary cognitive science, philosophy of science, and evolutionary theory. To be sure, the focus of this book is on understanding the complex and fascinating historical relationship between common sense and science stretching from Aristotle to Reid. To bring the story fully into the present, with a thorough examination of the relationship between common sense and the rise of modern evolutionary and physical theory, would require another volume of equal length. What readers should gain from the Epilogue is a sense of how the tradition of thought analyzed in this book played out in the nineteenth and twentieth centuries, how it relates to modern discussions of the relationship between common sense and science, and how it aligns with findings in contemporary cognitive science. As it happens, the common-sense tradition stands up very well in the light of current research into the mind and how it makes sense of the world. By the end, it is my hope that readers will be in a good position to come to their own conclusions about the enduring relationship between common sense and science, a relationship that remains of crucial importance to this day. One need look no further than current-day political leaders who get away with calling global warming a "hoax" or opponents of vaccines who are not even swayed by a worldwide pandemic that threatens their own lives and the lives of their loved ones, to find examples of the enduring rift between common sense and science. While there are many sources of the inability to fully grasp and act on scientific knowledge that has real practical benefit, not least among them is the rupture between common sense and science that was opened up in the early-modern era, and which persists to this day. If we can understand the cognitive roots of this rift, we may be better able to find ways to heal it.

Contextual factors—whether historical, intellectual, cultural, political, or social—are discussed as relevant throughout this study, as is the emergence of and usage of the term "common sense" in European social, cultural, and political discourse. Given the lack of a strict demarcation, for much of human history, between what we now call "science," on the one hand, and philosophy or philosophical discourse, on the other; and given the coherence and breadth of most philosophical systems, this study ranges widely over the history of the idea and philosophy of common sense and its relationship to science and scientific thinking. Much of early-modern philosophy was profoundly influenced by the New Science—and vice versa—and most of the thinkers treated in this study conceived of their enterprise as being in some measure "scientific," in the classical sense of systematic, rational, and/or empirical investigation. Until quite recently, what we now call "science" was typically referred to as "natural philosophy," and the term nicely captures the intermingling of natural science and philosophy throughout the early-modern period.

Finally, in charting the historical relationship between common sense and scientific inquiry, I allow the thinkers featured in this story to speak as much as possible, while making use of contemporary scholarship as needed, particularly when it comes to providing relevant historical and intellectual context. I ask that specialists of any one period or thinkers bear in mind the larger story that I am seeking to tell, and that it is simply impossible to pay attention to every nuance of any one philosophical system. My extensive use of primary sources provides readers with the evidence needed to assess the argument being advanced in this study. I believe the story told here helps us to understand some important features in the history of ideas, while also providing an overview of ancient, medieval, and early-modern intuitionist/innatist theories of mind and common sense. Every effort has been made to make the text readable without unduly sacrificing nuance and detail. Citations (in the form of end notes) along the way should give readers direction if they wish to explore any of the topics treated in this book in greater depth. Now, on to our story.

Chapter 1

COMMON SENSE AND SCIENTIFIC THINKING BEFORE COPERNICUS

Historians of philosophy suggest that philosophic reasoning arose in opposition to the common sense of the multitude.[1] The earliest Greek philosophers—lovers of wisdom and seekers of the truth that lay behind appearances—often cast themselves as those who went against popular opinions and everyday prejudices. The mysterious teachings of Heraclitus (540–480 BCE) and Parmenides (515–450 BCE) seemed to contradict what was known to everyone. Heraclitus suggested that very few would ever understand the *logos* or underlying order of the world, while Parmenides advanced cryptic arguments to the effect that change was only apparent, not real. Democritus (460–370 BCE) posited a world made up of tiny atoms that was quite different from the world as we experience it. In doing so, Democritus made a distinction between what would come to be known as "primary" and "secondary" qualities. The former are qualities inherent in matter itself, while the latter are the qualities of things as they appear to us. Such teachings, historians of philosophy suggest, "set philosophers against common sense in a most dramatic way."[2]

Plato (428–348 BCE) did much, in his writings, to distance philosophers from the common people and to differentiate true scientific knowledge (*epistêmê*) from the misguided and murky opinion (*doxa*) of the rabble. At the same time, however, he advanced the notion that knowledge of absolute truths is in some sense innate, requiring dialectical reasoning to be brought to light. Experience of the sensible world is misleading and cannot by itself provide us with scientific knowledge; there is an intuitive dimension to knowing. A variation of this idea (albeit with much more faith in sense perception) would form one of the core concepts of the modern philosophy of common sense.

According to Plato, Socrates held that the body, with its inaccurate and misleading senses, hinders knowledge.[3] Sight, touch, hearing, and taste provide only confused, relative measures of things, not their essence. True knowledge is abstract, universal, absolute, rational, and divorced from sense experience. The most that the multiplicity of sense experience can do is arouse perplexity and provoke thought, causing the mind to seek the essential unities that underlie the flux of experience.[4] It is the task of the philosopher to lead men on this upward journey out of the realm of sense experience to higher levels of wisdom.

Thus Socrates argues in the *Apology* that he loves the Athenians but must obey his God rather than men, never ceasing from the practice of philosophy, whose practical task was to make Athenians ashamed of putting so much stock in honor and money and reputation, and so little in "caring [...] about wisdom and truth and the greatest improvement of the soul."[5] Socrates famously likened himself to a large stinging fly on the rump of the Athenian horse, suggesting free meals in the Prytaneum (where Olympic heroes were fêted) as "punishment" for his activities.

In the *Republic* Socrates draws a distinction between the "sight-loving, art-loving, practical class" and philosophers. The former are "fond of fine tones and colours and forms and all the artificial products that are made out of them, but their mind is incapable of seeing or loving absolute beauty."[6] They and the rest of the multitude exist in the realm of "opinion," an intermediate zone between being and nonbeing where one sees only the changeable, *particular* exemplars of things like beauty.[7]

Philosophers, on the other hand, are the ones who seek (and perhaps even attain) the ideal. Such extraordinary souls are like tender plants that can grow to become healthy adults, or weeds, depending on the soil in which they are planted.[8] It would seem almost miraculous that there are any philosophers at all, given the corruptions of the public sphere. Individual sophists of the sort that Plato argued with in his dialogues, and who charged for their teachings in Greek homes and public squares, are not the real problem, according to Plato. The "public" is in fact the greatest sophist, educating the young and old according to its corrupt tastes. "Will any private training enable [a young man] to stand firm against the overwhelming flood of popular opinion?"[9] Sophists simply teach the opinions of the "motley multitude" back to itself.[10] For Socrates, "The worthy disciples of philosophy will be but a small remnant [...] Those who belong to this small class have tasted how sweet and blessed a possession philosophy is, and have also seen enough of the madness of the multitude." This remnant will be as men among beasts.[11]

The attainment of knowledge is possible only through dialectic, the process of philosophical argumentation by which one moves beyond the sensible to the ideal.[12] In the case of the Good, for example, one has to "abstract and define rationally the idea of [it] and [...] run the gauntlet of all objections, and [...] disprove them, not by appeals to opinion, but to absolute truth."[13] But how will we know the truth when we have found it? Plato's answer, in a nutshell, is that "learning is a process of recollection."[14] Intelligence of universals, Socrates says in the *Phaedrus*, is "the recollection of those things which our soul once saw while following God."[15] The process is illustrated in the *Meno*, where Socrates demonstrates how even a servant boy can discover the answers to a problem in geometry "out of his own head," through dialectical questioning. If he had such knowledge, but was not taught it, then his soul must have always possessed it.[16]

And so we have come full circle: the masses are misguided purveyors of opinion, sunken in the depravities of sense experience, yet the lowest among them is capable of rising above the confusion of appearances to the realm of *being*—with the help of the philosopher. The elitism that is deeply embedded in Plato's thought is thus tempered by the notion that built into each soul are the bare intuitions of the Forms (*eidê*) of absolute being. This view may perhaps appeal to those who argue that Plato was more of a democrat than has traditionally been thought,[17] without ignoring the critique of common sense and common opinion that is clearly evident throughout his writings.

* * *

Charting a different path, Plato's pupil Aristotle (384–322 BCE) championed common opinion and the assumptions and perceptions of everyday sense experience—what I will call common-sense experience[18]—as the foundation and starting point for scientific inquiry. If Plato distrusted common sense, Aristotle grounded his philosophy upon it. If Plato's philosopher was at best ambivalent about Athenian society, Aristotle took the *polis*, and the common knowledge that was its currency, as the locus of inquiry. For Aristotle, common-sense experience was the *basis* or *starting point* for science, rather than its foe. That did not mean, however, that scientific knowledge was the same thing as common knowledge or that Aristotle was not an empirical and rigorously independent thinker. His thought was in fact so comprehensive and synthetic that it became the template onto which medieval *scientia* was mapped.

Aristotelian thought was solidly based on common-sense experience and assumptions.[19] One starting point is language. In the *Categories*, for example, Aristotle assumes that critical analysis of the various meanings of words like "substance" can take us a good distance in understanding the basic structure of nature.[20] As Scott Atran puts it, for Aristotle "The primary objects of knowledge are those sensible things that *can* be named in ordinary speech (all emphases in quoted material are in the original)."[21] Second, throughout his writings Aristotle stresses the importance of assumed first principles or truths in all demonstrative reasoning.[22] As he says in the *Nicomachean Ethics*, "We must take great pains to state [first principles] definitely, since they have a great influence on what follows. For the beginning is thought to be more than half of the whole, and many of the questions we ask are cleared up by it."[23] In the *Physics*, after defining *what* "nature" is, Aristotle asserts it would be absurd to try to prove *that* nature exists: "for it is obvious that there are many things of this kind, and to prove what is obvious by what is not is the mark of a man who is unable to distinguish what is self-evident from what is not."[24] And the first sentence of the *Posterior Analytics* reads: "All instruction given or received by way of argument proceeds from pre-existent knowledge [...] The mathematical sciences and all other speculative disciplines are acquired in this way, and so are the two forms of dialectical reasoning, syllogistic and inductive [...] induction exhibiting the universal as implicit in the clearly known particular."[25]

But although we must start with self-evident first principles and with the way things appear to us, the universal, as Aristotle says, is only *implicit* in the "clearly known particular." Atran speaks of Aristotle's "extractive use of common-sense intuitions" in induction. "*Epagoge* [induction ...] is required to do much more than yield *de facto* universals. It must not only factor out the natural properties from the purely accidental aspects of a being (e.g. getting wet or wilted), but also the truly essential from the natural incidents of the common-sense type." Self-evident observations provide the basis for further logical analysis of the essential features of animal natures. A dog's canines, for example, allow it "to hunt, eat, grow and mature to look and behave as a dog is expected to; and it is because it has canines, as essential parts of its nature, to tear food and defend itself that it does not require the defensive horns of other animals nor the extra stomachs to complete mastication."[26]

It is not entirely clear from Aristotle's writings how we come to know first principles. In a much-disputed passage in the *Posterior Analytics*, he states that sense perceptions and *nous* (intellect or intuition), together with memory, combine to give us such truths.[27] In the *Nicomachean Ethics*, however, Aristotle is forthright on where we get our moral first principles: from a good upbringing.[28] In general, for Aristotle first principles emerge out of communal experience and common consent, and the Aristotelian notion of science (*epistêmê*) was concerned with revealing universal (or general) truths based on the commonly accepted facts of everyday lived experience.[29]

These "truths" of science are explanations of why things are necessarily so; what *causes* them to be as they are.[30] As is well known, Aristotle identified four different sorts of causes: the matter, the form, the mover or maker, and the end or "that for the sake of which" something is moved or made.[31] The "causes" of a statue would thus be the bronze of the statue, its form or archetype, the sculptor, and the purpose of the statue—for example, to commemorate a public figure. With this we have given a scientific account of a statue.

Drawing all of these elements together, Aristotle in the *Nicomachean Ethics* suggests that scientific knowledge involves demonstrating universal and necessary connections between things, proceeding via induction from known starting points to the generation of universal truths with the aid of syllogism.[32] Science is thus an activity of moving from the "givens" of common-sense experience, through inductive observations of particulars, to the general causal connections that bind everything together into one system of nature.

Aristotle's taxonomy reflects this model of science. Aristotle's science of biological classification provided the foundation for the development of modern taxonomy by grounding it in the "common sense appreciation of the living world shared not only by Aristotle and Linneaus, but by ordinary folk everywhere."[33] Aristotle's program involved "an unambiguous theoretical determination of the material natures of intuitive species. In this way, Aristotelian science renders ontological primacy to what is psychologically basic."[34]

As Atran demonstrates, there are certain universal patterns of "folkbiology" that are reflected in Aristotle's biological science. The folk categories of "generic specieme" (e.g., "woodpecker," "oak") and "life-form" (e.g., "bird," "tree") are universal and cross-cultural, although the specific exemplars within each category may vary.[35] Aristotle worked within this framework, reducing the common-sense "kinds" of living things to their essential parts, parts which could be "generalized to another by identity, by degree or by analogy." Disparate phenomenal kinds could then be related to one another, "thereby converting them into natural kinds that could be subsumed under unifying laws." The end result was to be "a taxonomy interlocking the essential attributes of different natures." In his biological inquiries, then, Aristotle "aimed to improve our understanding of the world as we ordinarily see it, and know it to be: not by refuting 'naive realism,' but by simplifying it."[36]

Aristotle's discussion of the place of the earth in the cosmos appeals in part to everyday observations of how heavy things move down and toward the center, light things up and away from the center. "The observed facts about earth are not only that it remains at the center, but that it also moves to the center," says Aristotle. The earth's immobility is further proven by the fact that there is no perceptible evidence that it moves, including the fact that objects do not move in any way that suggests they are participating in circular motion. And if the earth was in motion, then the fixed stars would appear to be "passing and turning" such that they wouldn't always rise and set in the same parts of the earth.[37]

Aristotle's dynamical theories are equally grounded in common-sense experience. Motion is the result of a proportional balance between force and resistance. "If [...] A the movent have moved B a distance C in a time D, then in the same time the same force A will move ½ B twice the distance C." However, experience shows that there are limitations to such rules of proportion. It is not necessarily the case that if you reduce the force by half, you move the object half as far. "Otherwise one man might move a ship, since both the motive power of the shiphaulers and the distance that they all cause the ship to traverse are divisible into as many parts as there are men."[38]

Motion is governed by rules of proportion that hold in the world of everyday experience. Thus although there are hints of the modern concept of inertia in Aristotle's writings, "it is simply the commonsensical empirical observation that objects once given a shove often tend to keep on moving—but certainly not (in idealized circumstances) indefinitely."[39] Common-sense experience tells us that matter resists changes to a state of rest, but not that it equally resists changes to a state of motion, and that motion and rest are relative states of being. As Isaac Newton was to put it, "Common parlance [...] assigns resistance to things at rest and impetus to ones in motion; but motion and rest as commonly conceived are distinguished one from the other only in their relative aspect, and things which are commonly regarded as at rest are not really so."[40] The seemingly absurd idea that we on earth are hurtling through space in a double motion requires mastering the counterintuitive idea that motion and rest are only relative states of being and applying everyday experiences of relative motion on earth to the solar system at large. For its part, Aristotle's theory of motion fully reflects the common-sense perspective.[41]

Generally speaking, then, while Aristotelian thought was in no way simpleminded, it was both methodologically and substantially in accord with common-sense experience. Scientific reasoning must begin with known and obvious (though not necessarily "popular") truths and assumptions, producing in turn substantive conclusions that are not in conflict with them.[42] And as if this were not enough to secure the foundations of science in common-sense experience, Aristotle espoused a psychology and perceptual theory that laid the foundation for the medieval notion of a *faculty* of common sense.

Aristotle's concept of an *aisthêsis koinê* or *dunamis koinê* is characteristically complex. The common sense is less a separate faculty than all of the special senses conceived as a whole. It is discussed mainly in reference to the perception of external qualities, rather than as an internal organ, although it is clear that for Aristotle its physical locus is in the heart. The common sense is a capacity shared by the special senses to perceive common sensibles—things like figure, magnitude, number, and motion—and to discriminate between the various qualities perceived by the special senses, as, for example, between white and sweet. It therefore gives us the ability to perceive the existence of different sorts of qualities in the same object and to know that we are perceiving the same thing in various ways. Charles Kahn suggests that Aristotelian common sense thus in some ways resembles our notion of consciousness, although it can't be separated, in a Cartesian sense, from sense perception.[43] As Aristotle puts it in *On Sleep and Waking*, "Each sense possesses something which is special and something which is common. Special to vision, for example, is seeing, special to the auditory sense is hearing, and similarly for each of the other others; but there is also a common power which accompanies them all, in virtue of which one perceives that one is seeing and hearing."[44] The common sense is thus accurately described, in the words of Pavel Gregoric, as a "higher-order perceptual power" of the soul.[45]

* * *

This "common power" was absorbed into medieval and early-modern understandings of the mind/brain complex, making the *sensus communis* "one of the most successful and resilient of Aristotelian notions."[46] The placement of the *sensus communis* in a specific location was one outcome of a process whereby a series of ancient writers fashioned a physiological model of human cognitive faculties pervaded by animal spirits and located in the ventricles of the brain.[47] By the Middle Ages, "In Avicenna's and most subsequent accounts of perception, common sense […] is located in the front ventricles, as is the faculty of representation [imagination/fantasy]."[48] The *sensus communis*, in the first step of cognition, initiates perception "by abstracting perceptible impressions from sensory input and channeling them into the imagination,"[49] where images are retained as short-term memories before being passed on to the rational faculties (middle ventricle) and then to long-term memory (rear ventricle). "Common sense" is thus given a physical location, but in general it "performs the perceptual function identified by Aristotle."[50]

The theologian Thomas Aquinas (1225–1274) situated Aristotle's doctrine of cognition squarely within the standard medieval physiological framework. Like Aristotle, Aquinas did not conceive of the *sensus communis* as the source of self-evident first principles, and he is a little bit murky on how it is that we know first principles. That we must build our scientific reasonings upon them, he has no doubt, but he also feels, with Aristotle, that the intellect is like "'a tablet on which nothing is written.'" He resolves the issue by saying that we know first principles by the agent intellect, which is the power to *realize*—or actualize—the self-evident truths via "habits of intellect." The ability to recognize first indemonstrable principles is simply a divinely given ability or power that springs into operation habitually (i.e., unconsciously) once we are presented with sense impressions. Thus although "Sensitive knowledge is not the entire cause of intellectual knowledge," "[the intellect's] first and principal objects are founded in sensible things." Aquinas' *sensus communis* is in the end very similar to that of Aristotle (whom he quotes constantly): a faculty of "discerning judgment" of proper sensibles that resembles a primitive sort of consciousness of sense experience. In a modification of Avicenna's model of the brain, Aquinas combines "phantasy" and "imagination" into a single storehouse of the forms received from the proper and common senses, forms which are then worked on by the estimative (or intellectual) powers before being passed on to the memorative powers.[51]

* * *

Thanks to Avicenna, Averroes, Aquinas, and a host of other medieval scholars, much of Aristotle's thought, including his natural philosophy, came to assume a dominant position in Western Europe between 1200 and 1450 and persisted, in various forms, well into the seventeenth century.[52] "As a group [...] medieval natural philosophers were convinced that Aristotle's metaphysics and natural philosophy, along with their corrections and additions, were sufficient to determine all that could be known about nature."[53] Aristotle was, in the words of Dante, "the Master of them that know."[54] Even as Aristotle's medieval disciples altered and expanded his natural philosophy, they "upheld its basic principles and remained faithful to its overall spirit."[55] The vast and elastic body of Aristotelian writings formed the basic content of Scholastic inquiry into nature, and as such Aristotelian conceptions of the scientific enterprise, and the role of common experience and indemonstrable first principles in that enterprise, became well established during the Middle Ages.[56] This is not to say that there was no great diversity in the thought of the period,[57] only that the main streams of Western scientific thought continued to flow, well into the Renaissance, in channels dug deep into the bedrock of common-sense experience.[58] This was all to change, however. The early-modern rejection of Aristotelianism and the rise of the New Science of Bacon, Kepler, Galileo, Boyle, and Newton displaced scientific thinking from its roots in common-sense experience, reinstating the rupture between the two that began with the Pre-Socratics and Plato. It is to this rupture between common sense and science that we now turn.

Chapter 2

THE CHALLENGE OF MODERN SCIENCE AND PHILOSOPHY

Modern science was founded in part on a distrust of ordinary sense experience and "appearances" in favor of corpuscular, idealized, and mathematical truths. The world of everyday experience needed to be reexamined, tested, transcended, put on the rack, and reduced (at least in theory) to invisible forces and minute particles in order to be understood. The writings of early-modern philosophers like Descartes, Hobbes, Locke, and Hume, who were deeply influenced by the discoveries and methodologies of modern science, exhibited this perspective, and the long-term effect was to place into doubt much that seemed to be self-evidently true and spontaneously known.

The first major event in this story was the advent of Copernicanism. There was no more significant and far-reaching challenge to the common-sense underpinnings of scientific thought and practice than the idea that the earth moves in a double motion around the sun—an idea which labored under the burden of contradicting Scripture as well.[1] The challenge to Scripture, and with it the medieval Christian worldview and power structure, often receives the most attention in discussions of the travails of Copernicanism. But as both Calvin and Galileo recognized, if there was a discrepancy between science and Scripture on astronomical questions, that was because the Scriptural writers had been addressing the common folk, on their own terms, about salvation and were not trying to teach them astronomy.[2] Thus although religious concerns and controversies were clearly central to the way in which Copernicanism and the new physics that went with it were received,[3] there can be little doubt that Copernicanism exemplified the challenges posed by modern science to common-sense experience, even as sense experience itself began to assume an authority that it had not held since Aristotle.

Copernicus himself realized the absurdity of his proposal, noting that it was "in opposition to the general opinion of mathematicians and almost in opposition to common sense [*communem sensum*] [that] I should dare to imagine some movement of the earth."[4] Here, Copernicus employs the term "common sense" to signify basic sense perceptions shared by everyone, as opposed to more abstruse mathematical knowledge. Copernicus's early Polish defender Barholomew Keckermann (d. 1609) argued that "'Everything indicates that the movement of the Earth as assumed by the greatly famed Copernicus must not be regarded as a realistic term but as an astronomical working term.'"[5] It was not until the early seventeenth century that Copernicus's proposal was taken to be anything more than a useful mathematical model of planetary motion and not until the later seventeenth century that it became widely accepted as such.[6]

Early responses to the idea that the earth was in motion were incredulous. The French jurist and political philosopher Jean Bodin pointed out that

> No one in his senses, or imbued with the slightest knowledge of physics, will ever think that the earth, heavy and unwieldy from its own weight and mass, staggers up and down around its own center and that of the sun; for at the slightest jar of the earth, we would see cities and fortresses, towns and mountains thrown down.[7]

A popular French poem first published in 1578 further set the agenda for Copernican apologists by pointing out that a moving earth implied a whole host of violations of common experience—arrows shot straight up would not fall back down to the archer, birds in flight would be catapulted eastward, cannonballs would recoil, and so on.[8]

It thus became necessary to begin translating the counterintuitive results of modern science into the terms of everyday sense experience. One of the most salient ways for Copernicans to do this was by using the analogy of being on a ship under sail. Copernicus himself first suggested the analogy by characteristically finding an ancient precedent—a line from the *Aeneid*: "'We sail out of the harbor, and the land and cities move away.'" "So," says Copernicus, "it can easily happen in the case of the movement of the Earth that the whole world should be believed to be moving in a circle."[9] In other words, just like it may appear that the *land* moves away from a ship while one is onboard, so it may only *seem* that planets and stars move in a circle about the earth. Galileo Galilei for his part famously used the ship analogy to argue for Copernicanism in his *Dialogue Concerning the Two Chief World Systems* (1632).[10] What better way to make sense of an absurd idea than to put it into terms that anyone who has traveled any distance will likely understand?

Galileo was well aware of the need to confront the challenge that Copernicanism—and the new physics of motion that went with it—posed to what seemed to be self-evident truths. In the *Dialogue* he has the Aristotelian Simplicius say that "Once you have denied the principles of the sciences and have cast doubt upon the most evident things, everybody knows that you may prove whatever you will, and maintain any paradox."[11] It was not going to be an easy task to overcome a scientific tradition rooted in common-sense experience. In its easygoing, conversational format, however, the *Dialogue* seeks to make the outlandish idea of an earth in motion amenable to everyday sense experience and to overcome the notion that violations of sense experience threatened the foundations of scientific knowledge.[12] The new hypothesis *could* be reconciled with common-sense experience, and hence all was not being thrown into doubt and confusion, as skeptics had been quick to point out.[13]

But although the Copernican hypothesis could be put into the terms of everyday sense experience, it grew out of other sources—reason and mathematics. As Johannes Kepler wrote in his *Epitome of Copernican Astronomy* (1618–21), the discipline that discloses the causes of things "shakes off the deceptions of eyesight, and carries the mind higher and farther, outside the boundaries of eyesight. Hence it should not be surprising to anyone that eyesight should learn from reason, that the pupil should learn something from his master."[14] John Wilkins expounded on this point at length in his Copernican tracts *The Discovery of a New World* (1638) and *Discourse Concerning a New Planet* (1640), adding a sociological dimension to the critique of common-sense experience. As he says in *The Discovery of a New World*, "You may as soon persuade some country peasants that the moon is made of green cheese [...] as that it is bigger than his cart-wheel, since both seem equally to contradict his sight, and he has not reason enough to lead him farther than his senses."[15] In both works Wilkins repeatedly denigrates the untutored sense experience of the common folk, and common opinions of all sorts, as being superficial and misleading. Only a more singular, impartial, private experience will lead to true knowledge. "And if in such an impartial enquiry, we chance to light upon a new way, and that which is besides the common road, this is neither our fault, nor our unhappiness [...] because it is rather a privilege to be the first in finding out such truths as are not discernible to every common eye."[16] And in a final blow to common sense, Wilkins, when discussing why it is that our senses don't tell us that the earth is moving, locates the root of the problem in the *sensus communis* itself, "because it apprehends the eye [...] to rest immovable, whilst it does not feel any effects of this motion in the body: as it is when a man is carried in a ship; so that sense is but an ill judge of natural secrets."[17]

The notion of a faculty of common sense was hence still alive—in the thought of Descartes and William Harvey, for example, or in Kepler's metaphor of the sun as a kind of cosmic *sensus communis*[18]—but it was not well,[19] and it would not be long before the old cognitive model was replaced by a new one, more in tune with the findings and methodologies of the New Science. Rather than talking about a single faculty spontaneously able to extract "common sensibles" from disparate sense data, modern philosophers working in the wake of the Scientific Revolution (and in the wake of the writings of John Locke in particular) increasingly saw the mind as a *tabula rasa* that took in data. This data normally assumed mental form as "ideas," which the mind then worked on in one way or another, normally via powers of "association" that connected ideas according to natural laws, to produce belief and understanding.[20] This was thought to be a more empirical and scientific account of the mind that assumed little (discussed further below).

* * *

The critique of the deliverances of common-sense experience and the *sensus communis* did not mean, of course, that sense experience was no longer important. Early-modern natural philosophy was becoming *more* rather than less empirical. But natural philosophers increasingly built their cases on particular and often recondite observations and *experiments*, rather than on that which is observed to happen in common *experience*, as Peter Dear has demonstrated.[21] What was needed, according to Wilkins, were "fresh experiments and new discoveries" rather than a "superficial knowledge of things" passed down from antiquity and reinforced by common experience.[22]

Sense experience, Francis Bacon argued at length, needed to be disciplined if it was to yield reliable knowledge.[23] "The assertion that the human senses are the measure of things is false; to the contrary, all perceptions, both of sense and mind, are relative to man, not to the universe."[24] While logicians had been happy to employ the "immediate perceptions of healthy senses" in their deliberations, Bacon argued that sense data and the first principles of the intellect require careful analysis and scrutiny, along with the aid of instruments and experimentation, in order to yield fruitful information. Traditional scientific inquiry had been far too quick to fly from the observation of "familiar things" to unfounded conclusions.[25] The axioms currently in use had come "from limited and common experience and the few particulars that occur most often, and are more or less made and stretched to fit them; so it is no surprise if they do not lead to new particulars."[26] What the sciences needed, in fact, was "a form of induction which takes experience apart and analyses it, and forms necessary conclusions on the basis of appropriate exclusions and rejections."[27]

A reformation of language was also needed, according to Bacon, since words themselves are forged in the crucible of common experience: "Words are mostly bestowed to suit the capacity of the common man, and they dissect things along the lines most obvious to the common understanding. And when a sharper understanding, or more careful observation, attempts to draw those lines more in accordance with nature, words resist."[28] For its part, the new Baconian science of induction "is not easy to get hold of, it cannot be picked up in passing [… and] it will not adapt itself to the common understanding except in its utility and effects." The New Science was to be *uncommon* and, if built on sense experience, aided by experiments and critical analysis of just what information the senses were, and were not, capable of delivering. Science needed to withdraw from the marketplace and go beyond the unreflective observations of common-sense experience if it wanted to "pass the antechambers of nature" and gain "access to the inner rooms."[29]

One result of this disciplining of the senses was that the traditional search for underlying essences and causes came to be replaced by the search for clearly observable "facts" and regularities, culminating in Newton's *hypotheses non fingo* (I do not feign hypotheses). As we have seen, *scientia* from Aristotle through the Middle Ages had for the most part involved demonstration of the necessary connections between things, proceeding via induction from known starting points to the generation of universal truths with the aid of syllogism. For Bacon too the aim was "to elicit the discovery of true causes and axioms from every kind of experience,"[30] but he began the process of elevating singular and hard-won particulars over universals as the starting point for natural philosophy,[31] and with Robert Boyle and his colleagues in the Royal Society the stress increasingly came to be placed on isolating discrete and recondite "facts" and behaviors, over the general hidden causes of sensible phenomena.[32]

It is true that the new "corpuscular" or "mechanical" philosophy of matter and motion was believed to be more intelligible than scholastic philosophy. As Boyle put it, "Men do so easily understand one another's meaning, when they talk of local motion, rest, bigness, shape, order, situation, and contexture of material substances."[33] But although mechanical explanations could provide an intelligible account of physical phenomena, such mechanisms were only taken to be plausible models. The point was *not* to assume that scientific knowledge begins with what everyone intuitively "knows" to be the case in nature—that is, with universal connections between things that simply announce themselves to us.[34] Such connections must be—and were—assumed to exist, but the credibility of natural knowledge came to be established through careful documentation of what had "happened" in the course of experimentation, through the discursive practices of gentlemen, and increasingly through mathematical methodologies and what Mary Poovey calls "gestural mathematics," rather than what self-evidently "happens" in nature.[35]

Galileo, working from within the Jesuit philosophical tradition, presented his novelties within the methodological framework of common experience,[36] but he grounded his science as much on mathematical certainties as on experiential ones, and he was diffident about the extent of our knowledge and skeptical of our ability to really "know" the nature of ultimate causes like gravity.[37] Thomas Hobbes similarly drew a distinction between the demonstrative certainty of geometry and the hypothetical, conjectural character of natural science,[38] and he even suggested that "very few" have the skill of scientific reasoning, as it is "not a native faculty, born with us."[39] By the time of Isaac Newton's *Principia* (1687), the universality of natural knowledge was no longer grounded in the common experience and intuitions of everyday life but in the combination of discretely observed and reported facts, often based on contrived experiments using special apparatuses, and the newly respectable "mathematical" method of analysis and synthesis.[40]

* * *

John Locke's philosophy of mind and science clearly reflected the early-modern turn in natural philosophy toward empiricism, nominalism, and the modest search for laws and regularities of behavior over the demonstration of universal and necessary truths built upon first principles of common-sense experience.[41] In Locke's view, too much had been assumed in the past, and now it was time to start over, to leave the "common road"[42] and assume as little as possible. In *An Essay Concerning Human Understanding* (1689) Locke vowed to "tether" human capacities to that which can be reliably known,[43] and to this end he worked hard to demolish the notion "that there are certain principles, both speculative and practical [...] agreed upon by all mankind [...] which [men] bring into the world with them."[44]

According to Locke, while it may be true that we assent to some basic self-evident propositions, "intuitive" knowledge comes about via the capability to immediately perceive the agreement or disagreement of two ideas, and nothing more.[45] Since ideas mediate our knowledge of the physical world, *certain* knowledge is restricted to knowledge of our ideas, while knowledge of corporeal substances and necessary connections in nature is provisional at best: "Though causes work steadily, and effects flow constantly from them, yet their connexions and dependencies being not discoverable in our ideas, we can have but an experimental knowledge of them." Since "we can go no farther than particular experience informs us of matter of fact," it is lost labour to seek after "a perfect science of natural bodies."[46] In Locke's philosophy, natural knowledge has moved from the realm of common experience to particular experience and from the quest for certainty to probability. Universal, public intuitions and certainties about the external world are rejected in favor of particular, private, and provisional perceptions of the agreement or disagreement of ideas.

* * *

That a Lockean account of cognition could lead to a clash with common sense is exemplified in the thought of the philosopher and clergyman George Berkeley. In *Three Dialogues between Hylas and Philonous* (1713), Berkeley struggled to convince the reader that his rather puzzling ideas are "agreeable to common sense, and remote from scepticism."[47] It is a dizzying performance, with Berkeley arguing that his idealism—the notion that nothing at all "can exist independent of a mind"[48]—only *appears* to run counter to common sense. What is really absurd, according to Berkeley, is the everyday belief in the independent existence of matter. Once a person understands perception rightly, it will become clear that the notion of the separate, independent existence of an unthinking substratum of the objects of sense leads to a skepticism that runs counter to common sense and, indeed, Scripture. It is in fact *Berkeley* who is of a "vulgar cast," a "simple," "plain" realist.[49] Berkeley protests too much, however, having Hylas repeatedly ask Philonous—Berkeley's stand-in—variants of the same basic question: "But be your opinion never so true, yet surely you will not deny it is shocking, and contrary to the common sense of men. Ask the fellow, whether yonder tree hath an existence out of his mind: what answer, think you, he would make?"[50] Thomas Reid for his part would have none of it, noting that "It is pleasant to observe the fruitless pains which Bishop Berkeley takes to shew, that his system of the non-existence of a material world did not contradict the sentiments of the vulgar, but those only of the philosophers."[51]

The paradox of Berkeley's *Dialogues* is that it takes a great deal of learned meditation to arrive at common sense. As Philonous admits, mental discipline—"time and pains"—is required to rid ourselves of the "prejudice" of believing in matter.[52] Berkeley believed he was able to accomplish this feat by modifying Locke's doctrine of primary and secondary qualities. Locke had developed the notion, introduced by Democritus and discussed by a variety of early-modern thinkers, that some qualities of bodies are "primary"—for example, shape, motion, and texture—while others are "secondary"—for example, colors, sounds, and tastes.[53] According to Locke, the primary qualities are inseparable from the bodies in which they inhere, while secondary qualities are "nothing in the objects themselves, but powers to produce various sensations in us by their primary qualities."[54] Both primary and secondary qualities are thus powers that produce corresponding "ideas" in us through our senses. Colors and tastes are ideas, produced in us by external bodies, that do not resemble or inhere in such bodies, while shape, motion, and texture are ideas that do resemble such bodies and "their patterns do really exist in the bodies themselves."[55]

What Berkeley did (and here he was preceded by the skeptic Pierre Bayle[56]) was to collapse the distinction between the two types of qualities. While Locke suggested that we could have ideas of primary qualities that really do exist in bodies, Berkeley argued that they too, just like colors and tastes, exist only as perceptions of our minds. Both primary and secondary qualities are mental realities, not physical ones, and hence all qualities are functions of the mind, and cannot exist independently of it. Thus our common-sense belief in the independent existence of material substances—for example, a glove apart from its qualities—was mistaken.[57] Since such a notion was clearly "repugnant to the universal sense of mankind,"[58] Berkeley valiantly (and one might say vainly) strove, in the *Dialogues*, to show that he was in fact involved in a "revolt from metaphysical notions to the plain dictates of nature and common sense."[59] But despite such protestations, Berkeley's bold opinion, Reid felt, would be interpreted by the unlearned "as the sign of a crazy intellect." Of all the opinions that had ever been advanced by philosophers, the idea "that there is no material world, seems the strangest, and the most apt to bring philosophy into ridicule with plain men who are guided by the dictates of nature and common sense."[60]

Writing in the early eighteenth century, Berkeley was clearly caught between what Hylas called "the modern way of explaining things" that seemed "so natural and intelligible"[61] and the dictates of common sense. That Berkeley repeatedly felt the need to justify his counterintuitive ideas in the court of common sense reflects the growing importance of the "commons" and indeed the notion of "common sense" in early-modern social and political life.[62] The term itself (as will be discussed in subsequent chapters) was coming into wider usage, particularly in Britain, in the early eighteenth century and was often identified with the "plain good sense" of the average Englishman. Thus just at the historical moment when politics and public life were becoming increasingly oriented toward growing middling classes, democratic politics, and the "sense of the people,"[63] the New Science and the philosophy that went with it began to diverge sharply from common-sense experience. Berkeley's *Dialogues* indicates that it was becoming increasingly difficult to advance ideas that violated common sense without attempting to show that such violations were only apparent and not real.

* * *

For his part, the eminent Scottish philosopher David Hume (1711–1776) openly acknowledged the modern scientific/philosophical departure from common-sense experience and developed its skeptical implications. This is not to suggest that there is not a constructive side to Hume's thought[64] or that his brand of empiricism has not become part of the standard methodology of modern science. But he himself characterized his philosophical quest as a solitary endeavor that led him to conclusions at odds with communal experience.[65] Hume was well aware that many of his notions went against everyday common-sense understandings and that no one, not even philosophers, allowed their own behavior to be affected in any significant way by such a viewpoint.[66] Hence in his *Essays* he was content to talk about "causes" as if they were real, external and knowable entities, and in *An Enquiry Concerning Human Understanding* (1748) he made use of the term "common sense" as a corrective to the excesses of religious enthusiasm and philosophical speculations.[67] In addition, Hume's moral theory can be seen as part of the anti-skeptical moral tradition that began in the early seventeenth century and was invigorated by the Third Earl of Shaftesbury and Hume as a "moral realist" whose "most explicit appeals to ultimate authority [in his moral theory] are to common sense or our common sentiments."[68]

Yet Hume's philosophy, and his metaphysics particularly, entailed a fundamental challenge to common-sense understandings of self and world and to the latent connections between common sense and science. As such, Hume served as a primary stimulus and foil in the rise of a coherent and clearly articulated philosophy of common sense.

Hume's challenge to common sense is clear, on the one hand, from his own programmatic statements. In his opening remarks to the *Abstract* of his *Treatise of Human Nature* (1740), for example, Hume states:

> Most of the philosophers of antiquity, who treated of human nature, have shown more of a delicacy of sentiment, a just sense of morals, or a greatness of soul, than a depth of reasoning and reflection. They content themselves with representing the common sense of mankind in the strongest lights, and with the best turn of thought and expression, without following out steadily a chain of propositions, or forming the several truths into a regular science.[69]

On the other hand, Hume also expressed a clear distrust of our intuitive perceptions of experience in ordinary life and as expressed by commoners. Thus in the *Treatise* itself (1739–40) Hume argues that the "vulgar" tend to "confound perceptions [... with the] objects [which they represent]," that is, they are unaware that their perceptions of an object are not identical with the object itself; in common life we are under an "illusion" about cause and effect, and the "vulgar" particularly are misguided in applying their rules; "superstition arises naturally and easily from the popular opinions of mankind"; and so on.[70] In his essay "Of Commerce" Hume makes a point about distinguishing between the understandings of the "common man" and "men of genius"—it is only the latter who are capable of "subtle and refined" reasonings about the "general course of things."[71] And elsewhere he argued that "popular opinion" was a "contagion" that needed to be managed so as not to be deleterious.[72] As Harvey Chisick points out, "It is ignorance more than anything else that characterizes the people for Hume. In [his works] the 'vulgar' are consistently opposed to philosophers, or those capable of a scientific and dispassionate understanding of things."[73]

Thus although Hume may have entertained a dialectical "philosophy of common life" that incorporated the perceptions of everyday life as the indispensable ground for skeptical inquiry, as David Livingston suggests, Livingston recognizes that Hume did not "claim to find in the order of common life a new foundation of true propositions on which philosophy could be built, as Thomas Reid and the Scottish school of common sense claimed to have found."[74] We shall return to the scientific challenge to common sense posed by Hume, and Reid's response, in Chapter 8. For the moment it will suffice to indicate the degree to which Hume's "science of man"[75] was in close accord with the intellectual tenor of modern science and philosophy, which entailed a fundamental shift away from conceiving scientific and philosophical knowledge as being grounded in common-sense experience.

* * *

Reid, besides emerging in a particular Scottish intellectual and religious context, was an inheritor of this wider European scientific and philosophical tradition. He read Bacon, Galileo, Boyle, and Newton, along with Descartes, Locke, Berkeley, and Hume. But despite this inheritance—or perhaps because of it—Reid's thought is marked by a profound conviction that modern philosophy had strayed dangerously far from the actual workings of the average human mind and from the language and understandings of common-sense experience. As we shall see, one of his fundamental tasks was thus to show the degree to which all forms of "scientific" thought and practice were rooted in what he called "principles of common sense." Religious, moral, educational, and general philosophical concerns were also clearly at the forefront of his thought, as they were in the thought of his precursors and followers. Yet Reid's thought was grounded in a thoroughgoing critique of the way in which modern "scientific" philosophy had drifted away from its moorings in common sense and into a self-defeating skepticism.

* * *

Modern science had not only undermined the perceptions of common-sense experience, and hence opened a chasm between common sense and scientific thought, but it had also given rise to a theory of the mind that challenged our ability to make sense of the world in ways that are recognizable to the untutored understanding. What came to be required was a *scientifically respectable* account of the mind that would show how we do in fact acquire knowledge of nature and nature's God in a way that did not lead to absurdity and skepticism. Such an account, by its very nature, would help to mend the growing early-modern rupture between common sense and science, while providing relief to religious sensibilities unsettled by the skeptical and materialistic currents of modern thought.

The notion of "common sense"—philosophical and otherwise—did not spring ex nihilo from Reid's head, however. Long before Reid published *An Inquiry into the Human Mind, On the Principles of Common Sense* in 1764, a variety of thinkers had already appealed to intuitive principles, notions, sentiments, truths, maxims, and to "common sense" itself, as they grappled with the skeptical and materialistic currents of modern thought. A number of Reid's central ideas had a long history, and it is to this intellectual tradition that we now turn.

Chapter 3

COMMON NOTIONS, *SENS COMMUN*: HERBERT OF CHERBURY AND RENÈ DESCARTES

The previous chapter focused on the ways in which early-modern science, and the philosophy that developed in its wake, came into conflict with common-sense experience and understandings. However, it would be a mistake to think that the modern philosophy of common sense was simply a reaction to the New Science; in fact, modern science was both a stimulus and a methodological resource for thinking about "principles" of common sense. And there were other intellectual currents in the early-modern period—including the rise of modern skepticism—that led a variety of thinkers who predated Reid to argue that human beings intuitively perceive certain notions, ideas, truths, or principles that condition our experience and make moral, scientific, and religious knowledge possible. This tradition of thought stretched from Herbert of Cherbury and René Descartes through the Cambridge Platonists and other English innatists to Henry Lee, G. W. Leibniz, the Third Earl of Shaftesbury, Claude Buffier, Frances Hutcheson, Joseph Butler, George Turnbull (Thomas Reid's teacher), and Henry Home (Lord Kames), Reid's friend and interlocutor.

While lines of influence between such thinkers are not always evident, and they did not all employ the term "common sense," they were responding to many of the same currents in modern thought, and in similar ways. Viewed against the background of such thinkers, the thought of Reid and his followers in the late eighteenth and early nineteenth centuries was a continuation and development of an existing, albeit loosely affiliated, tradition of modern thought. Reid was thus mistaken when he wrote in 1764 that the notion of "natural suggestions"—natural principles of mind like causality that are suggested by experience—had "entirely [...] escaped the notice of philosophers."[1] Like most philosophers, Reid was trying to establish himself as someone with something new to say, and he mainly had in mind those post-Lockean philosophers who subscribed to the notion that the main constituents of thought are "ideas" produced by experience and reflection. But those caveats aside, Reid was in this instance[2] ignoring a long tradition of modern thought, stretching back almost a century and a half before he wrote those words, that championed the existence of natural principles of our constitution, which in the course of experience suggest certain things, like the existence of causal connections in nature, that are not to be found in experience itself; and even his use of the term "natural suggestion" was not unique.[3]

In this chapter and the next I explore this tradition of thought from its early-seventeenth-century roots until the appearance of John Locke's *Essay Concerning Human Understanding* in 1689. The reason for this periodization is simple—Locke's *Essay* stands as an important caesura in the innatist response to modern skepticism and materialism. The *Essay* leveled a harsh and influential critique of the more naive versions of the doctrine of innate principles or ideas, while advancing a strict empiricism in accord with the methods and outlook of Britain's Royal Society, one of the most eminent scientific bodies of the early-modern period. Yet Locke had critics from the start who, while accepting his empiricism, reasserted the existence of natural first principles of our mental constitution that make experience and knowledge of all kinds possible. Although more simplistic conceptions of innate ideas or principles faded into the background after 1689, more subtle notions of "moral sense" and "common sense"—the latter culminating in Reid's "principles of common sense"—began to come into their own as terms of philosophical discourse. Thus if nearly all major works of philosophy after Locke bore the marks of his influence, not all philosophers were enamored with Locke's "way of ideas," and in many respects Locke's philosophy actually helped to stimulate deeper reflection on the intuitive and sentimental components of human knowledge. Readers may be surprised to learn of the degree to which such reflection was already prevalent in the seventeenth century, as early-modern thinkers such as Herbert of Cherbury and René Descartes—the main foci of this chapter—grappled with the skepticism that accompanied the rise of modern science.

* * *

With the rise of the New Science, an empirical and experimental approach to the study of human nature and society also began to emerge, coming to fruition in the "science of man" of the eighteenth-century Scottish Enlightenment.[4] Broadly speaking, this approach tended to marginalize the ineffable, spiritual dimensions of human experience—and of the world generally—in favor of identifiable behaviors and intelligible mechanisms. An entirely mechanistic view of human nature (we are simply machines that can think) and the world (everything is composed of matter in motion) began to loom as a threatening possibility to the religious mindset of the era, while our ability to know that world with any kind of certainty was also under attack by the rising currents of modern skepticism.

Emerging out of the religious controversies and debates of the Protestant Reformation and the subsequent wars of religion, early-modern skepticism drew from ancient sources, most importantly the writings of Sextus Empiricus (third century CE), to call into doubt human beings' ability to attain *certain* knowledge of anything. It was reinforced by (and helped to shape) the philosophical underpinnings of modern science, and it was informed by a new awareness, occasioned in part by the European voyages of discovery, of the diversity of beliefs and values held by human cultures. Michel Montaigne (1533–1592) was a seminal figure in the rise of modern skepticism, and his "Apology for Raimund Sebond," published in his *Essais*, exhibits three basic aspects of the emerging modern skeptical "crisis": theological, scientific, and cultural.[5]

The theological problem had to do with how one is rationally to justify any particular rule of faith. Since Montaigne felt that this couldn't be done, one must simply accept the authority of tradition and the religious faith given by God.[6] Montaigne's doubts about a rational basis for religion were extended, in the work of Isaac La Peyrére, Richard Simon, and Baruch Spinoza, to doubts about the divine origins and veracity of Scripture, and thereby to a general religious skepticism (and/or anti-Catholicism) that culminated in the work of Pierre Bayle and the French Enlightenment, and in the thought of David Hume, an avid reader of Bayle.[7]

The scientific problem, which according to Richard Popkin was the "most significant sceptical crisis precipitated by Montaigne," raised questions about "the reliability of sense knowledge, the truth of first principles, the criterion of rational knowledge, our inability to know anything except appearances, and our lack of any certain evidence of the existence or nature of the real world."[8] As Montaigne succinctly put it, "The senses themselves being full of uncertainty cannot decide the issue of our dispute. It will have to be Reason, then. But no Reason can be established except by another Reason. We retreat into infinity."[9] For Montaigne, the unreliability of scientific knowledge is indicated by the variety of plausible theories that have been put forward over time. For thousands of years most people believed that the skies were in motion; now Copernicus has provided a basis for supposing that it is the *earth* that is in motion. "For all we know, in a thousand years' time another opinion will overthrow them both." The best advice is to "'Believe nobody,' as the saying goes. 'Anyone can *say* anything.'"[10] Such doubts helped set the stage for the emergence of the "mitigated" skepticism of Marin Mersenne and Pierre Gassendi—along with much of early-modern scientific thinking—in which the sciences were understood to be hypothetical accounts of appearances rather than true descriptions of the essential natures of things.[11]

The cultural problem can also be called the problem of diversity: given the great variety of opinions and beliefs, both historically and geographically, how are we to justify the superiority of any one belief system over another? In the *Apology*, Montaigne repeatedly canvasses the wide range of beliefs and opinions held by a variety of thinkers, both ancient and modern, and he alludes to "natives of far off lands"[12] to highlight the diversity of beliefs and practices between cultures and to call into doubt the unique and superior nature of European culture.[13]

The moral and religious diversity of an expanding world presented a daunting challenge to European beliefs and helped to push thinkers away from traditional authorities and toward new intellectual foundations. In his *Essay* Locke appealed to a growing travel literature to argue that there are no innate moral rules or principles, since customs vary widely among peoples and men "have remorse in one place for doing or omitting that which others, in another place, think they merit by."[14] Locke's emphasis on the diversity of moral beliefs, and his efforts to disprove the existence of innate moral ideas or principles implanted by God, served to locate him squarely in the modern skeptical tradition,[15] even if his aim, as he put it, was to remove "some of the rubbish that lies in the way to knowledge" rather than to question our ability to gain any sort of knowledge at all.[16]

Indicating the way in which the skeptical appeal to diversity provoked appeals to common sense, Edward Stillingfleet, bishop of Worcester, argued in 1697 that Locke's account of primitive peoples "makes them not fit to be a standard for the Sense of Mankind, being a People so strangely bereft of common Sense, that they can hardly be reckoned among Mankind."[17] But already in 1646 the French skeptic La Mothe le Vayer had objected to this latter kind of argument, asking "But what is this common sense, what does it consist of? Is it a way of thinking common to a whole people?"[18] If so, then common sense is very mutable. Le Vayer argues that to appeal to common sense is only to defend one's own "folly" against that of others, and besides, "To think like everyone else is the safest means of thinking unwisely."[19] Appeals to reason don't fare much better. According to Le Vayer, reason "is a shameless courtesan who, covered with the mask of virtue, gives in disgracefully to all manner of opinion."[20] Clearly, moralists would need to face up to the challenge of cultural diversity and the vagaries of common opinion if they wished to argue for a universal moral consensus.[21]

* * *

The possibility of such a consensus was in fact vigorously defended early in the seventeenth century by Edward, Lord Herbert of Cherbury, in his book

De Veritate (*On Truth*; 1624, 1645). *De Veritate* ranks as the first significant modern exposition of innate principles of mind that make human experience and knowledge possible. Locke in fact mentions—and seeks to refute—Herbert in his *Essay*, in the same section in which he cites the worldwide diversity of beliefs as evidence that there are no innate practical principles of mind.[22] However, in keeping with the general reception of Herbert's ideas, Locke focused on Herbert's "common notions concerning religion" that were appended to the original text, while ignoring his main philosophical arguments.[23]

Written in idiosyncratic and murky Latin, *De Veritate* is known as a crucial founding text of Deism or "natural religion," an intellectual movement that downgraded Scripture and personal experience of a transcendent God in favor of rational and this-worldly marks of the divine. But although *De Veritate*'s general frame of reference was theological, as is to be expected of a European philosophical text written early in the seventeenth century, it in fact concentrates on explicating a theory of truth (or certainty) that rests on empirical appeal to a variety of innate principles, faculties, and "common notions" of mind, as well as the universal consent of all human beings, to make its case.

Herbert was a well-traveled man of action, an *uomo universale* of the Renaissance who played the lute, fought in the service of the Prince of Orange, wrote occasional verses, was fond of dueling, and served for a time as English Ambassador to France. The scion of a noble family from the Welsh Marches, Herbert lived for a number of years in Paris and spent a lot of time traveling throughout Europe. He made the acquaintance of French nobles and Italian scholars, of Gassendi and Descartes, and he even stayed with the renowned French scholar Isaac Casaubon. He was received by German princes and knighted by English kings, and he wrote a picaresque autobiography that extolled his exploits and the dashing figure he cut in the European *beau monde*. Caught between sides in the English Civil War, he made peace with the Parliamentarians (mainly to save his library), and then had to withstand a siege of Montgomery Castle by Royalist forces. And he wrote philosophical treatises, of which he was most proud: "Author of the book entitled *De Veritate*" was the sole accomplishment listed on his tombstone.[24]

If exposure to different cultures and peoples makes some individuals see only diversity, others—and Herbert was one of them—see commonalities. He writes confidently of common human (mainly European) beliefs and assumptions, normally without specific references, in the voice of one who has seen the world for himself. And he furthermore writes as the "hasty and choleric" person he understood himself to be.[25] Enough skeptical dithering, he says in so many words; there is a truth to be known and so let us put aside our petty differences and dig down to the fundamentals that we can all agree upon. It is in fact unsurprising that one of the first forays into a philosophy of common sense was made by a person fond of throwing down his glove and whose courage in battle bordered on rashness.[26] But if Herbert wants to furnish an answer to skeptics, he also concedes that there are many things that cannot be known with certainty and that there are many sources of error that can lead us astray. One of his primary purposes in writing, in fact, was to undermine all forms of dogmatic authority. A proper identification and ordering of "Common Notions"—a Stoic concept similar in some respects to what Reid would call "principles of common sense"—will, according to Herbert, "prevail over mysteries and faith and the arrogance of authority, and enable us to make a clean sweep of fables, error and obscurities."[27]

Herbert holds that truth can be attained when the appropriate "faculty" (or mode of apprehension[28]) conforms to its appropriate object, under the right conditions. According to Herbert, the human mind is endowed with a nearly limitless number of such modes of apprehension that correspond to the nearly limitless number of objects and qualities of the world. As such, there is an analogy or correspondence between the human microcosm and the larger macrocosm.[29] The conformity between faculty and object is confirmed by the stimulation of universally recognized Common Notions that are a mental form of natural instinct. In the course of our experience of the world, self-evident, universally held truths or principles are aroused in our minds, although they may go unrecognized. These Common Notions structure our experience and make possible the accumulation of further knowledge. Once we clearly recognize these instinctual principles, via introspection and ultimately through confirmation by universal consent, we are able discursively to arrive at other, more general truths (also called Common Notions). But rational analysis must be based on, and conform to, instinctual Common Notions, if it is not to go astray.[30]

Herbert emphasizes the point that "Truth, being a matter of conformity between objects and faculties, is highly conditional."[31] The task is carefully to identify the faculties, and the natural laws of conformity between faculties and objects, that is, the conditions under which things can be known to us.[32] The first condition of perception is that the object or thing to be known must "fall within our analogy," that is to say, it must fall within our range of perception.[33] Objects must be of a sufficient size, they must possess some distinguishing characteristic, they must persist in time, they must be related to one of our faculties, our sense organs have to be working properly, and so on, in order to be known with certainty.[34] As for the faculties themselves, Herbert designates four fundamental classes of faculties: natural instinct, which is that mode of apprehension that elicits and conforms to the Common Notions; internal apprehension; external apprehension; and discursive thought.[35]

Herbert also outlines four different levels of truth, including truth of thing, truth of appearance, and truth of concept, but the most important, comprehensive level of truth is "truth of intellect," the final level of conformity between things and our minds.[36] Truths of intellect are the "Common Notions found in all normal men; and by them, as though inspired from on high our minds are enabled to come to decisions concerning the events which take place upon the theatre of the world."[37] With the aid of Common Notions, the intellect can "decide whether our subjective faculties have accurate knowledge of the facts."[38] These "sacred principles are so far from being drawn from experience or observation that, without several of them, or at least one of them, we could have no experience at all nor be capable of observations." Thus if we had not been endowed with Common Notions to the end of examining into the nature of things, "we should never come to distinguish between things, or to grasp any general nature." By the same token, if we did not have any innate notions (and an analogous faculty) that distinguished between good and evil, "vacant forms, prodigies, and fearful images would pass meaninglessly and even dangerously before our minds."[39]

Herbert's goal is not to set forth new truths but rather to highlight "those [truths] which the reader habitually relies on without being aware of it."[40] "I term, then, those ideas especially Common which are shared by every man and can be excited by every type of object; such for example as that there is a first cause, an intermediate cause, and a final purpose of the world; that there is order, degree, change, etc., in things."[41] Other Common Notions include the principle of identity: "objects which stimulate our faculties in the same way are for us the same"[42] and the principle of non-contradiction: "principles that contradict each other cannot both be true."[43] Religion and God are Common Notions, in that no period or nation has been without them; yet specific beliefs, rites, ceremonies, and traditions vary.[44] Thus while the existence of a supreme God who ought to be worshipped is a Common Notion, and hence the basis for religious belief, any one particular revelation or Scripture must be carefully examined for marks of universal wisdom, otherwise it may be false.[45] In fact, all tradition and history, that is to say, all accounts of the past based on some authority, can only be probable, since they lie beyond the scope of our faculties.[46]

An important aspect of all Common Notions is that they tend to support self-preservation, not only of humans but of all "species, general classes and the Universe itself."[47] Without such an instinctual, universal law of nature all things would eventually be destroyed.[48] Aiding human beings in the quest for self-preservation and happiness is conscience, "the common sense of the inner senses." Conscience is a "tribunal of divine Providence" that helps to apply the Common Notions of shunning evil and seeking good to particular actions.[49] It resembles in many respects what would later come to be called the "moral sense," a faculty that intuitively judges the rightness or wrongness of actions. Herbert links conscience to the order of nature as a higher form of instinct that directs us toward eternal blessedness and salvation, the ultimate end of self-preservation and happiness.[50]

Herbert advances a number of criteria by which Common Notions can be recognized. First-order Common Notions come prior to discursive reason, they are independent and not derived from anything else, they receive universal consent, they are certain (no one can seriously doubt them), they are necessary for man's preservation, and they are brought into conformity (or perceived) immediately.[51] Of all these criteria, Herbert lays special emphasis on universal consent, which he calls "the final test of truth."[52] The prominent role given by Herbert to universal consent was criticized by Locke and has even received censure by modern commentators,[53] but for Herbert—as for the Stoics—the point is not that commonly held beliefs are necessarily true[54] but that the final test of a Common Notion is that it must receive universal assent.[55] As should be clear, Herbert advances a host of conditions and strictures on truth, and he appeals to individual perceptions of certainty as often as to universal consent, in establishing his doctrine. Right after saying that universal consent is the chief criterion of Common Notions, for example, he states that these notions present themselves as being *certain* to individuals: "As long as they are understood [by an individual] it is impossible to deny them."[56] One of these Common Notions is that all humans in all times and places share the same faculties.[57] Universal consent is thus a product of our faculties in two ways: it emerges from the agreement of the instinctual perceptions held by individuals, and its authority is itself derived from one of these perceptions (i.e., that we all share the same faculties). As such, universal consent is the ultimate confirmation of truths that begin at the level of individual minds. To *not* take universal consent into account would be to ignore self-evident truths and the testimony of all men in all ages, and hence leave oneself open to error.

In Herbert's view, he is narrowing and refining the sphere of truth, not opening it up to the vagaries of human belief. "It is not my intention here to fling open the door to any particular heresy or religious enthusiasm, nor to any type of intuition. I assert that truths which are universally demanded can easily be proved."[58] In Herbert's mind, by appealing to universal consent he is being scrupulously empirical and indeed objective, supplying a test by which anyone can measure their own, privately generated truths. That Herbert was willing to take this principle to its logical conclusion is indicated by his assertion that religious truths, including those of the Bible, are validated not by authority but by their conformity to our faculties and by universal consent—a momentous but logical step in an era riven by religious faction and controversy.[59]

In his autobiography, Herbert says that upon arriving in Rome, he made a point of telling the master of the English College, which was run by the Jesuits, that "the points agreed upon on both sides [Catholic and Protestant] are greater bonds of amity betwixt us, than the points disagreed on could break them," and that for his part he "loved everybody that was of a pious and virtuous life."[60] Herbert was clearly concerned to move beyond the differences of party, sect, and faction to the fundamental beliefs and values that were held in common. To do this required, ironically enough, a "hasty and choleric" temperament willing to offend all sides. Herbert vigorously asserted, against Calvinists, the reality of free will,[61] and against Catholics, that "the true Catholic Church is not supported on the inextricable confusion of oral and written tradition to which men have given their allegiance [...] The only Catholic and uniform Church is the doctrine of Common Notions which comprehends all places and all men."[62] For Herbert, "Those who enter the shrine of truth must leave their trinkets, in other words their opinions, at the entrance [...] They will find that everything is open or is revealed to perception as long as they do not approach it with prejudice."[63]

Herbert's irenic, anti-dogmatic approach to truth was grounded upon the assumption that there is a divinely ordained correspondence or analogy between human beings and the universe. Although he believes that we do not need to take this assertion on faith, Herbert asserts that God "has bestowed Common Notions upon men in all ages as media of His divine universal Providence."[64] This view, which permeates *De Veritate*, was a staple of the Platonic, Hermetic, and Stoic currents of Renaissance thought.[65] We can "make sense" of the world because we have been provided with the tools to do so: Common Notions "are, so to say, constituents of all and are derived from universal wisdom and imprinted on the soul by the dictates of nature itself."[66]

Herbert's ultimate aim was to delineate a method or theory by which everyone would be able to ascertain truth—or certainty—for themselves.[67] Truth is a collective enterprise and Herbert was less interested in supplying truths than to show how truth is generated and how individuals can ascertain it for themselves.

While it is difficult to provide a definitive assessment of Herbert's influence on subsequent thought, his main impact was on English and Continental religious writers and controversialists. His ideas on religion and religious common notions were widely known, if often at second hand, and he was eventually named the founder of Deism and attacked as a dangerous thinker who undermined the foundations of revealed religion.[68] Herbert does not appear to have exercised an impact on Thomas Reid, who does not mention Herbert in any of his writings, but later thinkers in the common-sense tradition came to identify Herbert as an important early forerunner of Reid and his followers.[69]

* * *

One very important early-modern thinker in direct contact with Herbert was the mathematician, philosopher, and scientific theorist René Descartes (1596–1650). Mersenne sent Descartes a copy of *De Veritate* in 1639. Descartes found some kind things to say about it, but said (privately) that Herbert's inquiry into truth was, in effect, pointless, since truth is a notion "so transcendently clear, that it is impossible to misunderstand it [...] Truth is self-evident; no analysis of it is possible, because it is assumed in any analysis."[70] If one didn't already know what truth is then there would be no way to know when one had found it. To agree with Herbert, one would need a criterion of truth that would validate the method of *De Veritate* for discovering it. Descartes seemed to imply that his own criterion, the *cogito*, was the missing element.[71]

Yet despite this criticism, Descartes wrote to Herbert that the book "has several maxims which seem to me so pious and so in conformity with common sense, that I would wish that they can be approved by orthodox theology."[72] He subsequently gave Herbert a first edition of his *Meditations*, and Herbert for his part began an English translation of Descartes' *Discourse on Method*. The two thinkers were in fact working on similar questions in the same intellectual atmosphere, and there are clear parallels in their thought when it came to innate first principles of mind that ground our experience and knowledge of the world.

Descartes plays a somewhat ambiguous role in our story. On the one hand, as a "conqueror of skepticism"[73] he located sources of certainty in the "good sense" and even "common sense" of the rational, but not necessarily learned, individual. He thereby laid the groundwork for a renewed conception of the links between scientific thought and the intellect of the average person. On the other hand, his strong distinction between mind and matter, and between mental certainty and the deceptions of sense experience, resulted in a thoroughgoing dualism between the inner world of human consciousness and the outer world of physical objects, the latter intelligible only through innately held ideas such as shape, extension, and motion. Such a dualism conflicted with common-sense perceptions of the reality and knowability of the physical world, and it made the external, public world of sense experience secondary to the private intuitions of individual mental experience.

Nonetheless, in attempting to provide new intellectual foundations for the sciences, Descartes grounded *scientia* in first principles and deductions intuited by the rational "good sense" (*bon sens*) of the average person, and in this regard Descartes' thought was friendly to common sense. Descartes equates good sense with "reason," "universal wisdom," and "the power of judging well and of distinguishing the true from the false."[74] This good sense is possessed by everyone and has little to do with book learning. As Descartes says in the *Discourse on Method* (1637), knowledge contained in books "never comes so close to the truth as the simple reasoning which a man of good sense naturally makes concerning whatever he comes across."[75] However mediocre his intelligence, anyone who has mastered the whole Cartesian method "may see that there are no paths closed to him that are open to others."[76]

In fact, unschooled people often make sounder judgments than those with great learning: "The learned are often inclined to be so clever that they find ways of blinding themselves even to facts which are self-evident and which every peasant knows."[77] Great souls are capable of great vices, while "those who proceed slowly can make greater progress," if they follow the right method.[78] And whoever discovers a truth knows as much about it as anyone else, as, for example, a child who has grasped a mathematical proposition.[79]

The sciences thus "are to be deduced from matters which are easily and highly accessible, and not those which are grand and obscure."[80] The common experiences of weaving or carpet-making, for example, are good activities to exercise our minds in instances of order.[81] In his unpublished dialogue *The Search for Truth by Means of the Natural Light*, Descartes' spokesman Eudoxus leads an "everyman" (Polyander) to recognize the basic tenets of his system, thus illustrating a method "which enables someone of average intelligence to discover for himself everything that the most subtle minds can devise."[82]

Descartes even refers to "common sense" (*sens commun*) now and then, as a synonym for "good sense." In the *Discourse on Method*, for example, he says that he is sure that his opinions "will be found to be so simple and so much in agreement with common sense as to appear less extraordinary and strange than any other views that people may hold on the same subjects."[83] In *The Passions of the Soul* (1649) Descartes suggests that it is humble people with "excellent common sense" who are most disposed to a state of wonder, and in *The Search for Truth* Eudoxus says "all we need for discovering the truth on the most difficult issues is, I think, common sense, to give it its ordinary name."[84] Here Descartes' felt need to introduce *sens commun* to his readers indicates the relative novelty of the term in early-modern philosophical discourse, as does his infrequent use of the term and his parallel use of other terms to suggest the same idea.[85]

In French dictionaries, "Bon Sens" appears early on as a synonym for common sense: already in 1606 we find in Jean Nicot's French-Latin dictionary the following entry: "*Homme de bon sens naturel* [man of natural good sense], Homo cordatus [wise or prudent man]." In 1627, in César Oudain's *Le Thresor des trois langues, espagnole, françoise, et italienne,* "Homme de bon sens" is discussed alongside "Sens naturel," which is translated into Spanish as "senso commun." In Guy Miege's 1688 French-English dictionary, *bon sens* and *sens commun* are discussed together as synonyms for "common Sense, or common Reason." And in 1690, in the *Dictionaire universel, Sens commun* is rendered as

> an interior power of the soul, that the Philosophers imagine to be in the brain, that receives all the species & images of the objects which strike the external senses […] Several believe that this is a superfluous supposition. *Sens commun* also means those general concepts with which men are born that makes them conceive things in the same way […] When one wants to accuse someone of being mindless, one says that they do not have common sense.

This passage juxtaposes the traditional *sensus communis*, standing for a specific faculty of the brain that unites data from the senses into a common perceptual awareness, alongside the emerging idea of *sens commun,* here defined as the everyday, commonly shared rationality of the average healthy adult.[86]

Descartes made use of both notions, locating the *Sensus Communis* in the pineal gland, while elevating the *bon sens* and *sens commun* of the untutored individual to the level of intuitive reason. His employment of both sets of terms is clearly mirrored in the *Dictionaire universel* and later on in the *Nouveau dictionnaire de l'Académe françoise* (1718). In this latter dictionary, after receiving the traditional Latin definition, *sens commun* is defined as "the faculty by which the majority of men judge things reasonably."[87] A similar definition appears in the nearly contemporaneous *Dictionnaire universel françois et latin.*[88] Diderot, in his *Encyclopédie,* authored the article on *bon sens,* for which *sens commun* appears as a synonym. *Bon sens* is "the measure of the judgment and intelligence a person can use to turn to his advantage common social undertakings."[89]

Whatever the term he used, Descartes made much of the role of unschooled intuition in perceiving the foundational truths that serve as the basis for scientific knowledge. As he says in the *Rules for the Direction of the Mind* (1628), "There are no paths to certain knowledge of the truth accessible to men save manifest intuition and necessary deduction […] It is clear that mental intuition extends to all these simple natures and to our knowledge of the necessary connections between them."[90] Descartes uses a variety of terms to describe the process of arriving at scientific knowledge, but the outline remains the same: scientific knowledge is achieved by intuiting foundational first truths and simple natures, and then proceeding via intuitively clear and distinct deductions to more abstruse truths and conclusions. "The whole of human knowledge [*scientia*] consists uniquely in our achieving a distinct perception of how all these simple natures contribute to the composition of other things."[91]

Ultimate certainty for Descartes is grounded in the bedrock of clear and distinct notions like "I am thinking, therefore I exist," or "two things equal to a third thing are equal to each other," or the idea of God: that is, in "innate ideas" or "common notions" or "simple natures" that are internal to the mind.[92] Such ideas can certainly have an external source—we are, for example, able to recognize, through the "innate light" of reason, such things as motions and figures—but our intuition of such simple natures involves our sense organs transmitting "something which [...] gives the mind occasion to form these ideas by means of the faculty innate to it."[93] The only immediate objects of sensory awareness are these ideas, and the sole sources of knowledge of material things are likewise these ideas.[94]

The senses themselves are misleading and provide only the bare outlines of knowledge. As Descartes puts it in *Principles of Philosophy* (1644), sensory perceptions "normally tell us of the benefit or harm that external bodies may do to [us], and do not, except occasionally and accidentally, show us what external bodies are like in themselves." If we bear this in mind we will lay aside "the preconceived opinions acquired from the senses" and "in this connection make use of the intellect alone, carefully attending to the ideas planted in it by nature."[95] Thus the intellect alone—not the combination of mind and body—is able to draw reliable conclusions from sense experience.[96] It is worth noting here that the Jesuit-educated Descartes incorporated the medieval model of the brain into his theory. According to Descartes, the purely spiritual power of intellect makes use of the figures or ideas that are received by the *sensus communis* and lodged in the imagination or memory in generating knowledge.[97] The *sensus communis* and imagination are located on the pineal gland, which for Descartes was the intersection between mind and body.[98]

Descartes, like Aristotle before him and Reid after him, warns against trying to give definitions of self-evident facts or truths; doing so will only obscure matters.[99] Like Herbert, he had come to believe that clearly and distinctly apprehended, self-evident truths were the only way to overcome skepticism, and this realization led him to identify truths to which all men could easily assent, thereby highlighting the original Aristotelian understanding of scientific knowledge as being grounded in the widest possible community of rational beings. But Descartes' version of this idea was profoundly individualistic and even solipsistic: the central actor in his philosophy is the solitary, meditative individual, carefully intuiting simple natures and their interconnections, and doubting everything else. And the foundational self-evident notions tend to be of the most slender kind: I am a thinking being and hence exist; God exists; two things equal to a third are equal to each other; and so on. Meanwhile the existence of external bodies can reasonably be doubted, the main guarantee for their existence being the clear conviction that God, who has given us the strong propensity to believe in the existence of corporeal things, is not a deceiver.[100]

Descartes found certainty in the clear and distinct ideas of the rational soul and in the existence of a non-deceiving God, while reducing the physical universe to a large machine working entirely on mechanical principles, knowable primarily through basic qualities of matter in motion. Remove God and disembodied spirits from the picture, and one is left with a purely mechanical universe of matter in motion—in other words, the world of Thomas Hobbes (1588–1679); and it is to innatist responses to his philosophical materialism—and the atheism that was felt to go along with it—that we now turn.

Chapter 4

HOBBES, LOCKE, AND INNATIST RESPONSES TO SKEPTICISM AND MATERIALISM

There are a variety of ways in which one can view the thought of Thomas Hobbes (1588–1679), but for our purposes he was important because he advanced a thoroughly mechanistic and materialistic view of human nature, informed by the ideas of Descartes, Galileo, and others,[1] that left little room for ineffable "spiritual" entities, free will, and innate first principles of mind that provide us with unmediated knowledge of an independent reality. Hobbes saw "man" as a purely material being driven by his passions and the need to survive (and hence as essentially solitary and selfish), and the world at large as being entirely constituted by matter in constant motion.

For Hobbes, "There is no conception in a man's mind, which hath not at first, totally, or by parts, been begotten upon the organs of Sense."[2] Sense is a "seeming" or "fancy" caused by an external motion or pressure on one of our organs, which gets communicated via nerves "and other strings" to the brain and heart and then produces a resistance or counterpressure that is directed outward to the original cause.[3] In step with the emerging scientific distinction between primary and secondary qualities, Hobbes asserts that colors and sounds and other such qualities are not in the bodies that produce them but rather in us; their external cause is only matter in motion.[4] The mind itself contains little, however, before experience. "There is no other act of man's mind, that I can remember, naturally planted in him, so, as to need no other thing, to the exercise of it, but to be born a man, and live with the use of his five Senses." All other faculties that seem proper to man "are acquired, and encreased by study and industry; and of most men learned by instruction, and discipline […] For besides Sense, and Thoughts, and the Trayne of Thoughts, the mind of man has no other motion."[5]

Belief and knowledge are hence based on our physical experience of the world. Science consists in our ability to order our sense experiences through the proper use of language: scientific knowledge is at root "knowledge of all the Consequences of names appertaining to the subject at hand."[6] Religious belief is based on the authority of men,[7] and the notions of good and evil emerge from human desires and aversions, "there being nothing simply and absolutely so."[8] There is no room here for inborn principles of mind that provide unmediated knowledge of either spirits or material things. "Truth" consists in the right ordering of the names that we give to the data of experience; it is an attribute of speech, not of things themselves.[9]

Whatever the content of his own religious beliefs,[10] Hobbes' materialism and frontal assault on the Anglican clerical establishment served to make "Hobbism" a byword for atheism, ethical relativism, and licentiousness in the seventeenth and eighteenth centuries.[11] Scholars argue there was in fact an "explosion" of atheism in Britain during the Restoration period, "largely confined to the upper classes and based primarily on the thought of Hobbes."[12] Daniel Scargill, a student at the University of Cambridge, made a sensational confession of atheism in 1669 that confirmed the fears of many divines that Hobbist atheists did in fact exist[13] and helped foster a growing anti-atheist literature that put forth a number of arguments for the existence of God and in support of the Christian faith, often buttressed by appeals to innate moral and religious principles.

* * *

According to John Yolton, in England the doctrine of innate knowledge "was held, in one form or another, to be necessary for religion and especially for morality from the early years of the [seventeenth] century right through to the end and into the beginning of the following century."[14] Variations of the doctrine grew out of the Aristotelian/scholastic and natural law traditions and "can be found in almost any pamphlet of the early part of the century dealing with morality, conscience, the existence of God, or natural law."[15] Writers like William Sclater (1611), John Bullokar (1616), and Richard Carpenter (1623) argued that moral and religious precepts were inscribed on man's inward conscience or soul and as such provided a solid foundation for morality.[16] Writing in 1646, John Bachiler stated that "the iniquity of the times, [has] so far corrupted the minds of some, that the very *innate and inbred principles of Nature* (especially about a *Deity*, the *sovereign welfare*, and the *Immortality of the Soule*) seeme in a manner to be quite *obliterated and extinct* in them."[17] As the century wore on, it became increasingly necessary to assert the existence and indispensability of such principles not only for moral and religious knowledge but for knowledge in general. Epistemological arguments often went hand-in-hand with moral and religious arguments, as in Matthew Hale's *The Primitive Origination of Mankind* (1677): "There are some truths so plain and evident, and open, that need not any process of ratiocination to evidence or evince them; they seem to be objected to the Intellective Nature when it is grown perfect and fit for intellectual operation, as the Objects of Light or Colour are objected to the Eye when it is open."[18]

Probably the most distinctive and important seventeenth-century British rebuttal to the growing skepticism, materialism, and scientific reductionism of the era is to be found in the works of the Cambridge Platonists. This was a group of seventeenth-century English moral and religious philosophers centered in Cambridge that espoused ideas which were, broadly speaking, "Platonic" in outlook and emphasis and as such advocated rational religion and the human ability to know attributes of the divine through reason. They were opposed to Calvinism, with its stress on the dogmatic and arbitrary will of God, as well as to the materialism of Hobbes and the impious trends of modern science and philosophy, Roman Catholicism, and the "enthusiasm" and "fanaticism" of religious sectarianism.

Benjamin Whichcote (1609–1683), Henry More (1614–1687), Ralph Cudworth (1617–1688), and John Smith (1616–1652) were some of the more prominent members of this "latitudinarian" movement of Anglican churchmen. They sought to defend belief in the existence of God and the immortality of the soul, free will, and the power of reason or the "natural light" to gain access to the divine. And an important feature of their general outlook was an insistence on the existence of innate, self-evident notions or principles of reason that give us moral and religious knowledge. The Cambridge Platonists therefore extended the moral and religious intuitionism present in the writings of Herbert of Cherbury and the pamphleteers studied by Yolton into a broad and recognizable movement.[19]

Henry More's *An Antidote Against Atheism* (1653, 1662) provides insights into the innatist aspect of Cambridge Platonism. The full title of the book is *An Antidote Against Atheism, or, An Appeal to the Natural Faculties of the Mind of Man, Whether There Be Not a God*. Although More, an avowed follower of Plato and Descartes, recounts wild stories of demon possession and miraculous phenomena to bolster his case against Hobbesian materialism, the general discussion had clearly shifted away from such things to careful analysis of man's natural faculties. According to More, "atheists" like Hobbes would have to be answered on their own terms—consequently More assumes the "plain shape of a mere *Naturalist*" in order to converse with them.[20] And an analysis of man's natural faculties reveals a number of innate ideas that cannot be reduced to the effects of matter in motion, including the "*Idea of a Being absolutely and fully Perfect*," natural remorse or conscience, and religious veneration.[21]

More begins by suggesting that the mind or soul of man is not an "Abrasa Tabula, *a Table-book in which nothing is writ*," but contains "*actuall knowledge*" or "an active sagacity in the Soul, or quick recollection, as it were, whereby some small business being hinted unto her, she runs out presently into a more clear and larger conception."[22] Here and elsewhere More makes clear that in speaking about "innate ideas" he does not mean static entities fixed in our minds like the "*Torches* or *Starres* in the *Firmament*," but rather an inborn faculty or capacity to recognize certain features of the world that are suggested by experience or some external impulse.[23] He uses the analogy of a musician able to sing a song upon hearing its first words: "So the *Mind* of Man being jogg'd and awakened by the impulses of outward Objects, is stirred up into a more full and clear conception of what was but imperfectly hinted to her from externall occasions."[24] There are a variety of notions that we have in mind, "which if we prove cannot be the Impresses of any material Object from without, it will necessarily follow that they are from the Soul her self within, and are the natural furniture of humane Understanding." Such mental "furniture" includes "*Cause, Effect, Whole* and *Part, Like* and *Unlike* [...] *Equality* and *Inequality* [...] and such like." These kinds of notions are not contained in the external objects that strike our senses, and thus must be the result of the Soul's "own active conception [...] whilst she takes notice of *external Objects*." In addition to such "Relative Ideas," however, there are several "complex notions" that the soul assents to "at the very first proposal," such as "*If you take Equall from equall, the Remainders are Equall.*"[25]

Although these arguments were mere prolegomena to More's main task, which was to demonstrate the existence and attributes of God, they are important (albeit undeveloped) milestones in the rise of a philosophy of common sense. The metaphor of "furniture" of the mind would be taken up by Reid's teacher George Turnbull, and the notion that the mind is an active interpreter of experience, containing faculties that are "awakened" to latent perceptions or truths, was an important component of common-sense philosophy, although philosophers writing in the wake of Locke were more careful about their use of the term "innate." Locke's critique of the term caused philosophers either to drop the term entirely or else clearly specify that by "innate" they meant principles or ideas that were latent in our minds before being called forth by experience—which is often called a "dispositional" theory of first principles. Yet as we have seen, in More's writings this dispositional theory was already clearly stated. The long-term trend was to speak more about the existence of active *faculties* or *senses* (including a moral sense and common sense) that are a function of our natural constitution and drop the notion of "innate" principles or ideas entirely.

More in fact employs the language of "sense" in *The Immortality of the Soul, So farre forth as It Is Demonstrable from the Knowledge of Nature and the Light of Reason* (1659, 1662). In his ongoing effort to rebut Hobbesian materialism, which reduced human actions to necessary physical causes, More (who himself coined the term "materialist") appeals to "that Faculty which we may call *Internal Sense* or *Common Notion*, found in all men that have not done violence to their own Nature," which attests to our freedom to choose to do good or evil. Everyone has such a "natural *Sense* or *Remorse of Conscience*" that is an undeniable witness to our moral freedom.[26]

The activity of the human brain cannot be reduced to merely mechanical motions. The "Common Sensorium," which Descartes located in the pineal gland, and which More holds to be the seat of consciousness, cannot possibly perform its various functions and still be material, unless such "matter" is itself spiritous.[27] As the soul's "Center of Perception" and "immediate Instrument for all manner of perceptions," the Common Sensorium gives rise to "*Secundae Notiones*"—the terms of logic and mathematics, as well as "the sense only of Sounds, Colours, of Hot, of Cold, and the like"—that cannot possibly be reducible to matter or to the names we give to them (as Hobbes had argued).[28] After much discussion, More locates the spiritous matter of the Common Sensorium—the animal spirits—in the fourth ventricle of the brain.[29] The soul, however, is not confined to the Common Sensorium but rather permeates the body.[30]

While More's Common Sensorium is not Reid's faculty of Common Sense, as a center of perception that produces immaterial notions as a result of experience, a certain similarity is evident. As in Descartes' writings, which More knew intimately, the traditional *sensus communis* is conceived to be the locus for intuitive perceptions and innate ideas. Hence there seems to be an evolutionary relationship between the *sensus communis*, as a faculty of the brain occupying a specific location, and the emerging notion of a faculty and/or principles of common sense that were a general feature of the human mind. Over time, the former would be dropped as a result of the empiricist project of focusing on identifiable mental phenomena and behaviors rather than on speculative accounts of brain function, but the latter would remain, "common sense" making an appearance even in the writings of arch empiricists like Locke and Hume.

Although More claimed to approach his subject as a "naturalist," the purpose of his works was to demonstrate the existence of *super*natural, spiritual entities like God and the human soul. His empiricism included a strong tendency to accept reports of spiritual phenomena as reliable evidence for the existence of such entities, as, for example, in his recounting the story of a Spanish abbottess who possessed powers that allowed her to know what was occurring in distant locales, to receive Eucharistic wafers transported by angels through walls and to be lifted 3–4 cubits (about five feet) off the ground.[31] To those steeped in the methods of the New Science, such credulity would seem to call into doubt the whole innatist enterprise—just look where acceptance of innate principles can lead! As Locke put it, "If it be the privilege of innate principles to be received upon their own authority, without examination, I know not what may not be believed, or how anyone's principles can be questioned."[32] Doctrines derived from the superstitions of a nursemaid "may, by length of time and consent of neighbors, grow up to the dignity of principles in religion or morality" that are taken for "unquestionable, self-evident, and innate truths."[33]

Locke was not the first to offer a critique of early-modern innatism, however. In 1666 Samuel Parker, in *A Free and Impartial Censure of the Platonick Philosophie*, argued that appeal to "first and fundamental Principles" leads the minds of men away "from the native Evidence of plain and palpable Truths," causing them to "ground all their knowledge upon nice and subtle speculations." Parker, a fellow of the Royal Society, objected to the Platonic philosophy primarily because of its method. Whereas Platonists start with the assumption that "God has hang'd a multitude of [...] little Pictures of himself and all his creatures in every man's understanding," members of the Royal Society have discarded all particular hypotheses "and wholly addicted themselves to exact Experiments and Observations" that allow them to furnish the world with a complete history of nature and "lay firm and solid foundations to erect Hypotheses upon."[34]

In this view, the New Science had made significant progress by eschewing accepted truths and the dictates of common sense in favor of observation, experimentation, and mathematical calculation. Parker's attack, published in the year of Newton's discovery of the calculus, was a harbinger of things to come. There was to be no going back to the old ways and supposed truths of the past, when new methods were becoming available that offered a firmer, if shallower, foundation for knowledge. A scientific disciplining of the mind would cleanse it of unwarranted assumptions and opinions, as well as unwarranted skepticism, according to Locke.[35]

* * *

But "scientific" skepticism and materialism were to remain the bugbears of early-modern thought. Whether philosophers argued in favor of innate ideas, self-evident first principles, common notions, first truths, maxims, axioms, or common sense itself, all were concerned to assert the human ability to make rational, indubitable sense of the world and to combat atheistic materialism. Defense of religious belief and moral principles were often the animating concerns, but as we have seen in Chapter 3, anti-skeptics like Herbert and Descartes were concerned with more than merely defending orthodoxy.

Certainty was a good in itself, and first principles of mind were seen to be necessary for the production of all types of truths, whether religious, moral, or scientific. And belief in a rational order that pervades the universe, and is intelligible to a mind created to respond to that order in very specific ways, meant that a defense of common sense and reason tended to go hand in hand. Herbert's incipient Deism, Descartes' overt rationalism, and More's Platonism all point to the close connection between "reason" and the various terms that Reid would later gather together under the umbrella of "principles of common sense." Reid himself called common sense "the first-born" or "first degree" of reason.[36]

Such an identification of common sense and reason is evident in Robert Ferguson's *The Interest of Reason in Religion* (1675), a text that anticipates some of the arguments later made by Reid. Like Reid, Ferguson (1637–1714) was from the Aberdeen area and may have attended the university at Aberdeen, before becoming a Presbyterian minister. Unlike Reid, however, Ferguson moved to England and went on to lead a colorful and notorious life that surpassed even that of Herbert of Cherbury for sheer chutzpah. Known as "the plotter," Ferguson was involved in nearly every antigovernment plot that took place in the last third of the seventeenth century. Ferguson took part in efforts to oust Charles II and to keep James II from becoming king. Forced into hiding in Holland a couple of times in the early 1680s, he triumphantly returned with William of Orange in 1688, only to turn against him to become a Tory and Jacobite in the 1690s, now participating in plots *against* William. Imprisoned several times, Ferguson always managed to escape execution, surviving to write, in the last years of his life, books chronicling the many plots and rebellions in which he had been involved. Widely perceived as an incorrigible traitor, it is hard to imagine a more antiauthoritarian character than Ferguson: no matter who was in power, Ferguson was against him.

Ferguson wrote many pamphlets and more than a few books. *The Interest of Reason in Religion*, published in 1675, defends human reason as "*the Candle of the Lord*" that provides the light needed to ascertain "the Authentickness and Sense of Revelation."[37] Second, however, "Reason is taken Metonymically for common *Maxims*, or principles whose Truth is inviolable" and are "so connate to Sense and Reason [...] that upon their bare Representation they are universally assented to."

I do not say that we are brought forth with a List and Scroll of *Axioms* formally imprinted upon our Faculties; but I say that we are furnished with such Powers, upon the first Exercise of which about such things without any Harangues of Discourse, or previous Ratiocinations, we cannot without doing Violence to our Rational Nature, but pay them an Assent.

> Those Truths whether *Logical, Moral, Physical*, or *Mathematical*; Whether *General* (because of their Universal Influence upon all Disciplines) or *Particular* (from their being confined in their Use to some one Science) are justly stiled Natural, being founded on the Nature of God, the Essences of things, and the intrinsical rectitude of the Rational Faculty. These are the Foundations and Measures of all Science, Knowledge, and Discourse; being in themselves certain and incontestable.[38]

To the degree that men partake of the same reasonable nature, "the certainty of these Principles is Universal." Such principles include the principles of identity and causality—*"every Effect supposeth it's cause*; and many such like."[39]

Both science and faith build upon such principles, forming the basis for what Ferguson calls "Acquired Principles," similar in some respects to Reid's "acquired perceptions."[40] Such principles "are discovered by a Chain of Ratiocinations, and their Verity established by a Harangue of Inductions."[41] Acquired principles, if properly derived from unquestionable premises, are no less true than self-evident principles, although more remote from view. Problems in both science and faith occur when, due to our fallen nature, "we do often prevaricate in making Deductions and Inferences from self-evident, and universal Maxims, and thereupon establish Mistaken and Erroneous Consequences, as the Principles of Truth and Reason." The problem is not reason, but the use that men with all their "Lusts [...] Passions [... and] sensual Appetites" make of it.[42]

Ferguson occasionally makes reference to Aristotle and other ancient writers in his treatise, but he also indicates that he is conversant with the debates and figures of modern philosophy, mentioning Descartes, More, Gassendi, and Parker, among others, when considering whether or not we have an innate idea of God. For his part, Ferguson steers a middle way by saying that "I know no *Idea's* formally *Innate*; what we commonly call so, are the Results of the Exercise of our Reason. The Notion of God is not otherwise inbred, then that the Soul is furnished with such a Natural Sagacity, that upon the Exercise of her rational Powers, she is infallibly led to the Acknowledgment of a Deity."[43]

Ferguson's basic argument is that we are equipped with a "Natural Light, and such common Principles which all men assent to"[44] that provide the basis for all forms of knowledge, illuminating not only the natural world but Revelation as well. But Ferguson does not "make Natural Light the positive Measure of things Divine." Rather, reason is a kind of gatekeeper that "hinder[s] the entring of Contradictions and Irrational Fancies, disguised under the Name of Sacred Mysteries" into the counsels of faith.[45] Ferguson identifies this "natural light" with common sense, as when he says that error and heresy emerge when "men [fill] Religion with Opinions that are contrary to common Sense and Natural Light." For Ferguson, doctrinal problems occur when "*Dogm's* Repugnant to Right Reason" are admitted, as when "The first Hereticks that troubled the Christian Church, under pretence of teaching Mysteries, overthrew common sense and did violence to the Universal Uniform and perpetual Light of Mankind."[46]

Here, mundane common sense is held up as an arbiter of religious faith: if religious teachings contradict plain common sense and natural reason, they should be rejected. The deistic tendency is obvious, as is the rebellious Ferguson's similarity to the hasty and choleric Herbert. The appeal to universal principles of mind, and/or a natural sagacity to discern such principles, permits one to defy established authority, whether religious or political, in favor of the universal and perhaps even subversive truths that are self-evident to everyone, including the common folk, who are thereby enlisted as partisans in whatever battle one is currently fighting. Ferguson's early use of the term thus indicates that there is no necessary connection between appeals to common sense and conservative politics, as is often assumed today.[47] On the other hand, one can also discern hints of a modern-day anti-vaxxer or global warming denier: "so-called experts tell us X (vaccines keep populations healthy, the earth is warming due to human causes) but anyone can see that Y (some children die after being vaccinated, cold weather happens even in Spring) is the case." Common sense can serve as a powerful court of appeal against "authorities" of all kinds, including scientific authority; hence, science ignores the court of common sense at its own peril.

Ferguson further suggests that *moral* principles are also a result of our natural sagacity. "As soon as we come to have the use of our Intellectual Faculties, we are forced to acknowledge some things Good, and other things Evil. There is an Unalterable Congruity betwixt some Acts and our reasonable Souls, and an Unchangeable Incongruity betwixt them and others." Furthermore, "We find ourselves possessed of a Faculty necessarily reflecting on it's own Acts; and passing a Judgment upon it self in all it does"—in other words, conscience.[48]

Although Ferguson writes more as a rationalist than as a partisan of common sense, he lays out a general (albeit rather sketchy) philosophical perspective that was widely held at the time[49] and would later be taken up by respondents to Locke, Berkeley, and Hume. His thought builds on the rationalism of Descartes and the deism and general philosophical outlook of Herbert, although it is unclear whether Ferguson had read Herbert. He sounds at points much like the Cambridge Platonists, but he is clearly contemptuous of the kind of spiritual credulity found in More's writings and wants a strictly rational religion, going so far as to assert that Jesus "spake illogically" when he spoke of the Eucharistic bread as being his body.[50] Ferguson's main target in 1675 appears to be Roman Catholicism, but he also mentions Lutheranism in the same breath as containing equally absurd creeds.[51] The natural light of reason and common sense seemed to have been Ferguson's main allies, as he did battle with the authorities and creeds of his day.

* * *

Although "common sense" thus began to appear in English philosophical discourse by the later seventeenth century, it did not merit a separate entry in seventeenth-century English dictionaries.[52] However, the Oxford English Dictionary (OED) identifies literary examples from the sixteenth and seventeenth centuries, the earliest appearing in 1535: "I am suer T[indale] is not so farre besydis comon sencis as to saye the dead bodye hereth cristis voyce." In examples from 1561 and 1602, common sense is identified with "natural wit," and Shakespeare used the term to mean ordinary or untutored perception. The OED also identifies uses of the term in 1663 and 1695 where it is employed to signify "the general sense, feeling, or judgement of mankind, or of a community."[53]

In the 1611 and 1650 editions of Randle Cotgrave's *A Dictionarie of the French and English Tongues*, the French *bon sens* is translated not as common sense but as a form of wisdom.[54] By 1677, however, in Guy Miege's *A New Dictionary of French and English*, "Il n'a pas les sens commun" is translated as *"he hath not common sense."* And then in the 1688 edition of the same dictionary, "le Sens Commun" is now defined as "the rational and intellectual light that most people are born with, the common Sense, or common reason, a Mans Understanding."[55] As noted by Rosenfeld, Edward Phillips, in a *A New World of Words: Or, Universal English Dictionary* (London, 1706), "defines common sense as 'those general Notions that arise in the Minds of Men, by which they know, or apprehend things after the same manner,' a direct translation of the definition of *le bon sens* offered by Antoine Furetière in his *Dictionnaire universel* of 1690 and a standard for the rest of the eighteenth century."[56] Samuel Johnson's *Dictionary of the English Language*, first published in 1755, still did not feature a separate entry for common sense, although in the entry for "sense" common sense is identified with "natural light, and reason."[57]

John Locke had identified common sense with the natural light of reason long before Johnson's *Dictionary*, but he believed that any talk about "common maxims or principles" of mind that are held to be in any sense of the term "innate" allowed for unthinking acceptance of existing creeds and authorities. For Locke, "Principles must be examined" and not simply received upon their own authority.[58] Such examination revealed that nearly all supposed self-evident principles of mind could be accounted for by our "natural faculties [...] without the help of any innate impressions, and [we] may arrive at certainty without any such original notions or principles."[59]

As for the argument that universal consent indicates the existence of innate principles of the human mind, Locke felt that "universal consent proves nothing innate"—all it proves is that there are some things all men agree upon.[60] Besides, if such innate truths were imprinted on the soul, then children and idiots would perceive them, but they do not.[61] And as we have seen, Locke argues that if there were innately held *moral* principles, then there wouldn't be so much moral diversity in the world.[62] Diversity in belief also argues against there being an innate idea of God; only "right use of [our] reason" can give us that idea.[63] Just as God was not obliged to build bridges and houses for us, he was not obliged to implant ready-made ideas of himself and of morality in our minds, when we may attain the same through the proper use of our natural abilities and faculties.[64] Locke thus argues that there *are* innate or natural powers, tendencies, desires, and aversions, "but this makes nothing for innate characters on the mind, which are to be the principles of knowledge, regulating our practice."[65]

In seeking to demolish the notion of innate ideas or principles of mind, Locke provides an alternative view of how the mind works that in essence separates the objects of the understanding—"ideas" received from experience—from the mind's inherent powers, capabilities, and operations.[66] On this model, intuitive knowledge arises when the mind perceives the agreement or disagreement of two ideas "immediately by themselves, without the intervention of any other [ideas]."[67] Self-evident truths or propositions are therefore those to which we immediately give assent, once the terms are known, that is, once they are learned through experience.[68] Thus the principle of identity becomes self-evident as a result of "getting" and then comparing the ideas of concrete things (and the terms used to express them). As a result of this practical exercise, the mind eventually comes to appreciate the abstract, self-evident "principle of identity."[69]

Similarly, our ideas of cause and effect arise from experience: "In the notice that our senses take of the constant vicissitudes of things, we cannot but observe that several particular both qualities and substances begin to exist; and that they receive this their existence from the due application and operation of some other being. From this observation we get our ideas of cause and effect."[70] The principles of identity and cause and effect are thus known as a result of experience, not because of being imprinted on the mind from birth, even if only in latent form.

Philosophers stretching from Herbert of Cherbury to Thomas Reid, however, insisted that without some expectation and understanding of notions like causality, prior to experience, we would have no grounds for identifying their referents within the flux of experience. Causality is an *interpretation* of phenomena not contained in phenomena. Hume, as we shall see, developed this viewpoint to its ultimate skeptical conclusion, but he offered a way out of the dilemma by suggesting that experience of the constant conjunction of ideas creates a habitual sentiment or *feeling* of their connectedness. This "solution" to the problem of causality would prove unsatisfying, however, to those who believed that there is a rational order of things and that knowledge—moral, religious, scientific—cannot and should not be reduced to habit and feeling.

Locke for his part was relatively untroubled by such questions. Experience, added to our rational powers of comparing ideas, appeared to be enough to provide us with more complex notions and principles. Our mental operations as described by Locke thus mirror those of the good Baconian natural philosopher, producing knowledge as a result of observation, experimentation (repeated experiences), and induction from particulars. If there was to be any agreement on the meaning of terms, what was needed was an acknowledged way of thinking that took nothing for granted except that which can be experienced in the same way by everyone. "I can only appeal to men's own unprejudiced experience and observation," says Locke, to verify "whether [my principles] be true or no."[71] This was not the "common experience" of Aristotle as discussed in Chapter 1, with its overtones of commonly held truths emerging out of shared everyday experience. This was rather the highly particularized experience of the Royal Society, wherein particular facts (Locke's "simple ideas") entered, via particular unprejudiced individuals (Gentlemen), into a common stock of verifiable yet provisional truths, when not amenable to mathematical demonstration.[72]

Yet Locke himself appealed to "common sense" on occasion, indicating the ineluctable role of spontaneous perceptions in human judgment and knowledge, or at least the difficulty of expunging such terms from one's vocabulary. "He would be thought void of common sense who asked on the one side, or on the other side, when to give a reason, why it is impossible for the same thing to be, and not to be. It carries its own light and evidence with it, and needs no other proof."[73] Here and elsewhere common sense appears as that power which is able spontaneously to distinguish between one thing (idea) and another.[74] In some passages, Locke links common sense with reason and "the very principles of all [men's] knowledge."[75] Locke would no doubt argue that the identification with reason is a defensible use of the term—but "the very principles of men's knowledge"? Locke in fact comes very close, in that passage, to affirming the existence of principles of common sense that are intuited by the average healthy adult. Nevertheless, as we have seen, Locke would argue that such principles are derived via reflection on experience. But the question remains, how are we to know that *these* principles are any better than the "fancies" that they are said to contradict? Locke makes it quite clear that "any absurdity" may be taken to be an innate principle, because of the power of custom, education, and opinion, the power of experience, in other words.[76] The answer appears to be that we just have an "intuitive certainty" about some principles.[77] Locke would seem to be right—we must be intuitively certain of at least some principles prior to experience. For example, if principles are to be derived from our ongoing experience of the world, we must have an intuitive certainty that the world is an orderly place. Without such an assumption, there would be no basis for trusting that we can learn anything useful from repeated experiences or from simultaneous experiences occurring in different places. The inherent circularity of such an argument is clear: in order to get principles, we need already to have them—or at least some. It is unsurprising, then, that in crafting an epistemology that rejects all forms of innate knowledge, Locke appeals to common sense and intuitive certainty from time to time, despite his best efforts to ban all forms of innatism from his philosophy.

Locke's critique of the doctrine of innate principles of mind was very influential, however, and helped lay to rest more "naive" versions of the doctrine, in which certain precepts were imprinted by God onto the soul at birth. Locke's thought was widely disseminated both in England and abroad in the years after the publication of his *Essay* in 1689, and his overall empirical emphasis and method was widely accepted as an important advance in the theory of knowledge and the science of man.[78] No one could afford to ignore Locke's account of human understanding and his rejection of innate principles of mind. By 1728 Ephraim Chambers could state with confidence, in his *Cyclopædia*, that "the Doctrine of *Innate Ideas* is abundantly confuted by Mr. Locke." According to Chambers, philosophers of "the best Note [...] deny the Reality of any Innate, or *Common Notions*; urging, that the Mind does not need any actual *Notions* to prepare it to think, but that an innate Faculty of Thinking may suffice."[79]

Innate ideas, harkening all the way back to Platonic forms (*eidê*) that are brought with us into the world, were no longer plausible philosophical notions, but the nature of the faculties or powers capable of intuiting first principles was still an open question, as was the relationship between the internal world of ideas and the external world of sense perception. As we shall see, if Locke had vanquished "innate ideas," it would not be long before others would suggest that *his* doctrine of ideas was itself problematic on a variety of grounds. The terms of the debate had shifted, but the debate was far from over. In fact, the stage was now set for the emergence of a truly "modern" philosophy of common sense that would begin to bridge the widening chasm between common sense and science.

Chapter 5

COMMON SENSE IN EARLY EIGHTEENTH-CENTURY THOUGHT

The traditional narrative of modern philosophy emphasizes the triumph, over the course of the eighteenth century, of Lockean empiricism over various forms of rationalism. But thinking in terms of a dichotomy between "rationalism" and "empiricism" obscures more than it clarifies. While it is true that thinkers following in the wake of Locke's *Essay* were forced to take the role of sense experience much more seriously than before, a salient feature of eighteenth-century philosophy was its exploration of just how the mind is able to gain understanding from experience. And this exploration often led philosophers to sharpen and refine innatist arguments, rather than reject them entirely, as Locke had done. Locke's attack on innatism thus helped to stimulate the emergence of a mature philosophy of common sense.

Although the term "common sense" had occasionally appeared in seventeenth-century philosophical discourse, it tended to be invoked in a rhetorical manner, as when Robert Ferguson stated that the first heretics of the Christian Church "overthrew common sense, and did violence to the Universal Uniform and perpetual Light of Mankind" or when John Ray said that the common sense of mankind tells them that animals suffer.[1] But already in such appeals, the term functioned as a synonym for practical reason or commonly held principles or truths. During the eighteenth century, as "common sense" began to be widely used in the service of a growing egalitarian and democratic mindset,[2] thinkers started to accord it more philosophical weight and substance. It was becoming increasingly important to reconcile elite knowledge with common understanding, particularly as middling classes assumed ever greater prominence in cultural and political life, and thus it should come as no surprise that the term assumed a new role in the vocabulary of philosophers during the eighteenth century.

* * *

Locke's *Essay* received many critical responses after its publication in 1689, and his rejection of innate principles was a central point of contention. As

Yolton states, many people "found Locke's rejection of innate ideas, his treatment of conscience, and his dissociation of the law of nature from the inward principles commonly supposed to be inherent in man dangerously challenging to the established morality and to revealed religion."[3] A number of clergymen responded to Locke by defending the doctrine of innate moral and religious principles and highlighting the skeptical implications of Locke's "way of ideas."[4] A generally shared criticism of the latter was that it opened an unbridgeable gulf between the human mind and the external world, a point forcefully argued in John Sergeant's *Solid Philosophy Asserted, against the Fancies of the Ideists* (1697). According to Sergeant, who was an ardent if unorthodox Catholic priest, the "*Ideists* [...] do presently begin to imagine that those *Ideas* have got rid of the *Thing*, and hover in the Air (as it were) *a-loof* from it, as a little sort of shining Entities." The "solid Nature of the Thing" is neglected in favor of "meer *Material* Resemblances, or Phantasms," which can no more produce true knowledge than can a looking glass. For Sergeant, the word "notions" is better suited to "express distinctly those Solid Materials, by the composition of which the structure of *Science* is to be raised."[5]

Just as Locke's "ideas" derived from experience seemed to depart from the real world of solid things, his terminology seemed to depart from common language. Edward Stillingfleet, bishop of Worcester and a vigorous critic of Locke, claimed that he did not object to the use of the term "idea," "provided it be used in a common sense [...] for I am for no new affected *Terms* which are apt to carry Mens Minds out of the way."[6] Stillingfleet argued that Locke's way of accounting for principles like causality using the theory of ideas made a simple thing complicated: "For is not any Man who understands the meaning of plain Words satisfied that nothing can produce itself?"[7] Locke had suggested that his terminology was in line with common sense, and Stillingfleet took pains to show that this was not the case. In his view, Locke's way of ideas could not "give a satisfactory Account as to the Existence of the plainest Objects of Sense." That is because "*no Idea proves the Existence of the thing without it self, no more than the Picture of a Man proves his Being.*" And if we cannot be assured of our basic rational perceptions—including the existence of external objects and a first cause of all that exists—then "*Assurance of Faith upon Divine Revelation*" becomes even less likely: "Now I appeal to any Man of Sense, whether the finding the Uncertainty of his own Principles which he went upon in Point of Reason, doth not weaken the Credibility of [...] fundamental Articles [...] of *Faith?*"[8] There must be some principles, antecedent to revelation, that ground our belief in what we read in the Bible. For Stillingfleet, the proper foundations of certainty lay not in the comparison of ideas but in "*general Principles of Reason*" or "the Self-Evidence which attends the immediate Perception of our own Acts."[9]

Epistemology was thus closely linked to theology, and so it became neces-sary, in view of Locke's new account of human understanding, for clergymen like Stillingfleet to refine or at least restate the innatist theory of knowledge. Although Stillingfleet appealed to common sense only sporadically and without philosophical precision,[10] his response to Locke pointed the way toward a philosophy that grounded not only religious but scientific knowledge in first principles spontaneously intuited by the average person.

* * *

An important early philosophical response to Locke's attack on innate principles that includes appeals to common sense was Henry Lee's *Anti-Scepticism: Or, Notes Upon Each Chapter of Mr. Locke's Essay Concerning Humane Understanding* (1702), a book that has been described as "one of the most able and detailed [contemporary] assessments of [...] Locke's *Essay.*"[11] Lee had been a fellow of Emmanuel College, Cambridge, before becoming rector of Tichmarsh in Northamptonshire. He was concerned to outline the ways in which Locke's philosophy led to skepticism and to champion the existence of innate, self-evident principles of mind—what he sometimes calls "common principles of reason"—that provide a sound basis for knowledge and truth. In critiquing Locke's "way of ideas," Lee advanced ideas and perspectives that would later be shared by philosophers of common sense, and he invoked "common sense" in a way that prefigured their use of the term. Lee's book, coming at a point in time halfway between the writings of Herbert of Cherbury and Thomas Reid, highlights the existence of a continuous philosophical tradition that responded to the skeptical implications of modern thought by appealing to universally shared first principles or truths that could not be proven or learned from experience.

Lee opens his book by suggesting that it had become the common fashion to doubt "whether there be any such thing as real Truth; for the receiv'd Maximes of all Mankind, which used to be the Touchstone, by which to try it, must be tried themselves, and in the mean time are to be reckon'd purely arti-ficial, and wholly owing to the powerful Influence of Custom and Education." While other philosophers were remaking not only natural science but the very foundations of church and state, Lee was "contented with the honest and humble Design of speaking the common Language of Philosophy" in order to alert others, including "the Vulgar," to the skeptical implications of the new, Lockean philosophy.[12]

Lee thus diagnoses the rise of modern philosophy as a departure from the truths of common-sense experience, and he reinforces his own position as a defender of received wisdom by occasional appeals to common sense.

Lee wonders, for example, whether it "is not contrary to common sense" to suppose that we cannot know certain truths because we do not have a conscious idea of their terms.[13] Locke had made conscious awareness of an idea or principle a condition for calling it "innate," but Lee argued that we call propositions innate

> not because we are born with an actual Notion of all the Particulars in our Minds, but with a natural Facility to know them, as soon as the Things imply'd in those Word are presented to the Understanding, and a natural and unavoidable Determination to judge them true, as soon as we know the Things themselves or the Words by which they are signif'd to others.[14]

Lee believed that Locke's theory of ideas would inevitably lead to a thoroughgoing skepticism that was at odds with truths known by everyone. The theory of ideas could not guarantee the existence of external objects, just as it could not account for fundamental principles of nature like causality, since it set up an unbridgeable gulf between the mind and the external world:

> Any man may easily satisfie himself that his Hand is bigger than one of his Fingers [...] merely by comparing their Ideas; but then you must observe, that the Mind all the while supposes the real Existence of the external Causes or Objects of those Ideas [...] If you offer to suppose the real Existence of any thing out of the Mind itself, then you go beyond your Ideas; for they are as wholly within the Mind, as the things themselves are without it, and therefore have no Connexion in Nature with each other.[15]

If you try to prove the existence of external objects with abstract and general "Ideas of Cause or Effect," this will not work because we do not have any abstract general ideas "answering to those general or common words." Causes and effects are known to us as real substantial entities, and the mind connects them together without recourse to explicit "ideas" of causality.[16]

The only way out of the dilemma, then, is to accept that there are indeed "such general Truths, as the Mind in all its Reasonings, Arguings and Judgments always and necessarily supposes true, as it does the truth of its own Faculties. They are self-evident, need no proof or indeed are capable of any; the Mind assents to them immediately following the very Perception of the Relation there is between any of the particulars."[17] Examples include the principles of identity and noncontradiction, cause and effect, and the whole is equal to the sum of its parts. "Cannot any one that is or has been Twenty years

deaf and dumb, have the Notion of Cause and Effect, without ever understanding, or so much as hearing any of those two words, by the feeling of Heat
from fire, or Light from the Sun, or a hundred other ways?"[18]

Just as we assume certain speculative principles, so we also take for granted
practical principles—moral rules. They "are as generally and readily assented
to as any self-evident speculative Principles, and all Men too would be govern'd
by them as constantly as the others are judg'd true, but that some unreason-
able Passion overpowers, for the present, those Inclinations which Nature has
planted in their Minds."[19] Murder is the same in England as in Peru, and just
because moral laws are not always observed does not mean they do not exist.[20]
Every man knows to preserve his life and that of his offspring just as well as he
knows that he exists and has legs and hands; to promote and preserve peace
"is as obvious a Truth, as that a strait Line is the shortest;" and to do as one
would be done by "is as plain as that Equal added to Equal makes the Whole
equal."[21]

Just as the mind unavoidably assumes certain speculative as well as practical principles, it assumes the truth of our senses and other faculties. And "as
we unavoidably suppose our own, so we do without any Proof, that the Senses
and Faculties of other intelligent Beings in the like Cases are true also: For if
we did not suppose that, neither they cou'd prove any thing to us, or we to
them."[22] Meaningful communication, in other words, requires that we assume
the veracity of our own faculties as well as those of others. A consequence of
assuming the veracity of others' faculties is that when we are in doubt about
something, we by "the same unavoidable necessity of our constitution resort
to those common Sentiments of all Mankind, as the surest Principles into
which to resolve the reason of our belief of any Proposition which our own
Senses or Observations could not discover."[23] Those natural assumptions that
help us to make sense of the world and to communicate with others thus lead
us to look to the general principles held by all mankind as the surest way to
justify our beliefs, "presuming inevitably, that such Universal Sentiments are
the Impressions of the wise and undeceiving Author of our Natures." If we
assume that God has created us as a part of a coherent natural order, then
we may trust in those "common Principles of Reason" shared by everyone.[24]

The assumption that God equipped human beings with the faculties or
notions needed to make sense of the world was, as we have seen, a staple
of innatist theories from the beginning. This "providential naturalism" (as it
is sometimes called) would continue to form the backdrop for philosophies
of common sense into the nineteenth century, albeit with a tendency to discern the existence of God and a providential order from reflection on nature
and experience rather than from Scripture. Lee, for example, argues that the
notion of God is "innate" in the sense that it is formed in our minds, without

teaching or arguments, "by the efficacy of natural Causes operating upon us, and the unavoidable Observation of such Effects as can proceed from no less or other Cause, than such as we all mean by the Word God."[25] The notion of God is thus the product of a capacity or disposition shared by all humans to intuit certain things as a result of experience.[26]

If any proposition can be resolved into such self-evident truths, it is supposed to be sufficiently proved, "because when so resolv'd they appear to be agreeable to the general Sense of Mankind, and need no farther Proof or Evidence."[27] Although Lee's explicit appeals to "common sense" tended to be fairly rhetorical,[28] his philosophical ideas very clearly anticipate those of later philosophers of common sense.

* * *

Another significant early response to Locke was Gottfried Wilhelm Leibniz's *New Essays on Human Understanding*, completed by the time Locke died in 1704 but not published until 1765. Locke's *Essay* had been reviewed in French-language journals published in Holland, and Leibniz had gotten hold of a French translation of the book published in 1700. In his own "new essays" on human understanding, the German philosopher counters Locke's attack on innate principles of mind in ways that also foreshadow later responses by common-sense philosophers, although he does not appeal to "common sense" itself—not surprising, given the fact that the term came into currency later in Germany than in Britain and France, due in part to quite different social and political conditions.[29] Leibniz argues, contrary to Locke, that there must be innate ideas or principles of mind that, in concert with sense experience, produce universal and necessary truths—scientific knowledge in the traditional sense. Sense experience, while necessary for all our actual knowledge, is not sufficient for establishing its universality, since the senses only provide particular instances of general truths. "However often one experiences instances of a universal truth, one could never know inductively that it would always hold unless one knew through reason that it was necessary."[30] Here Leibniz neatly expresses what later came to be known as the "problem of induction," which asks how are we able to induce general laws from particular events if all that we actually observe are particular events. Leibniz says that we have "a disposition, an aptitude, a preformation, which determines our soul and brings it about that [universal truths] are derivable from it."[31] In other words, we are equipped with innate principles of mind that are capable of binding together individually observed events into universal truths.[32] Leibniz uses a variety of terms to talk about these innate principles, including eternal laws of reason in the soul, instinctive truths, fundamental assumptions or things taken for granted, common notions, imprinted items of knowledge, and innate principles, ideas, and maxims.[33] The specific terms did not matter so much as the basic idea that sense experience alone cannot give us universal and necessary truths: our minds must add *something* to experience that enables us to attain the truths of math and geometry, logic, metaphysics, and ethics, along with those of natural theology and natural jurisprudence.[34]

For Leibniz, innate truths include the principles of identity and noncontradiction, being, unity, substance, duration, change, "and hosts of other objects of our intellectual ideas." Such "in-built principles of the sciences and of reasoning" are a species of natural instinct that we employ without knowing the reasons for them. The idea of God is in the depths of our souls, "and some of God's eternal laws are engraved there in an even more legible way, through a kind of instinct." Such practical principles include moral concepts like justice and temperance.[35]

Leibniz stresses the point that these innate principles are necessary but not sufficient conditions for knowledge. They are latent until called forth by experience, just as veins in a block of marble that mark out the shape of Hercules would be in a sense "innate" in the marble, even though labor would be needed to bring this potential Hercules into being. "This is how ideas and truths are innate in us—as inclinations, dispositions, tendencies, or natural potentialities, and not as actualities."[36] General principles serve as the "inner core and mortar" of our thoughts. "Even if we give no thought to them, they are necessary for thought, as muscles and tendons are for walking."[37]

Leibniz acknowledges Locke's insight that talk of innate principles can be a cover for intellectual laziness, and he champions the rigorous analysis of all ideas and truths, even primary ones.[38] Yet Leibniz recognizes that one cannot provide proofs for innate principles, since they underlie our thought processes, and he suggests that "it is enough that [innate principles] can be discovered within us by dint of attention: the senses give the occasion, and the results of experiments also serve to corroborate reason, somewhat as checks in arithmetic help us to avoid errors of calculation in long chains of reasoning." It is in this sense that "The nature of things and the nature of the mind work together."[39]

For Leibniz, the rational mind, when employed properly and with due attention to experience, is uniquely able to gain natural knowledge. In reply to Locke's assertion that "natural philosophy is not capable of being made into a science," Leibniz argues that while the whole of natural philosophy will never be a perfect science, "still we shall be able to have some science of nature, and indeed we have some samples of it already," such as the science of magnetology. "From a few assumptions grounded in experience we can demonstrate by rigorous inference a large number of phenomena which do in fact occur in the way we see to be implied by reason."[40] The art of discovering the causes of phenomena is thus a kind of deciphering: "an inspired guess often provides a generous short-cut." While Robert Boyle was a gifted practitioner of Baconian rules of experimentation, he "does rather spend too long on drawing from countless fine experiments no conclusion except one which he could have adopted as a principle, namely, that everything in nature takes place mechanically—a principle which can be made certain by reason alone, and never by experiments, however many of them one conducts."[41]

Leibniz's critique of Boyle's exhaustive experimentalism was clearly of a piece with his answer to Locke's wholesale rejection of innate ideas. Just as it was philosophically unacceptable to suppose that experience alone can provide universal truths, so it was scientifically naive to suppose that experimentation by itself can lead us to principles of nature. But Leibniz the great mathematician assumes that "science" involves establishing universal and necessary truths along the lines of mathematics and geometry. As we have seen, this view was being supplanted, in some circles, by a more skeptical and diffident view of scientific knowledge. Nonetheless, Leibniz was simply raising objections to the new empiricism that had been raised before, and would be raised again by philosophers like Claude Buffier and Thomas Reid, who would add a sociological dimension to innatist arguments. For these thinkers, modern skepticism and idealism were problematic not least because they appeared to lead to a debilitating rupture between higher, "scientific" thinking and common-sense experience.

* * *

That such a rupture would come to be perceived as problematic was due in large part to fundamental changes underway in European social, cultural, and political life, and nowhere were these changes more apparent than in Britain. Over the course of the eighteenth century, merchants, traders, lawyers, and other professionals became increasingly important arbiters of British public life, as an extra-parliamentary political culture began to take shape in London as well as other urban centers throughout Britain. The "commons," with its own religious, moral, and cultural sensibilities, was on the rise, and the "sense of the people" and "common sense" were becoming important touchstones in British politics, promising as they did to serve as a new basis for consensus in a society tired of political faction and religious strife.[42] According to Sophia Rosenfeld, "In the decades that immediately followed the revolutionary settlement, the ordinary sense of the ordinary man in ordinary circumstances was envisioned as a respectable, trustworthy, and superior standard for judgment in such seemingly disparate arenas as religion, ethics, aesthetic taste, justice, and politics."[43]

The term "common sense" thus began to appear early in the century in periodicals like *The Tatler* (1709–11) and *The Spectator* (1711–12) as a practical, down-to-earth faculty of deliberation that put the lie to pedantry, abstract theorizing, infidelity, elitism, self-regard, the vagaries of fashion, and fool-ishness of all kinds. These two periodicals, which were published by Richard Steele and Joseph Addison, were short-lived but significant expressions of a new literary sensibility in Britain. The conversational tenor of these papers replicated the democratic, sociable ambience of London coffeehouses, giving scope to the "plain common sense" or "good sense" of the average Englishman, whatever his (or her) social background. Writing in *The Tatler* in 1709, for example, Addison suggests that "a few solemn blockheads" want to appear wiser than everyone else, propagate infidelity and "extirpate common sense" by publishing such "crude conceptions" as that the soul is not immortal, or men are no better than brutes. Here, common sense is iden-tified with "those schemes of thinking, which conduce to the happiness and perfection of human nature" and which had characterized the British nation in the past.[44] Steele, for his part, railed in *The Spectator* against freethinkers in similar terms: "They can think as wildly as they talk and act, and will not endure that their wit should be controlled by such formal things as decency and common sense. Deduction, coherency, consistency, and all the rules of reason they accordingly disdain, as too precise and mechanical for men of a liberal education."[45] In both of these examples, common sense appears as a synonym for generally accepted understandings of morality and religion and for practical reason.

In the April 30, 1710, edition of *The Tatler*, Addison contrasts common sense with pedantry: "men of deep learning without common sense" will on the one hand dismiss beautiful poems composed by their contemporaries, but on the other hand "will lock themselves up in their studies for a twelve-month together, to correct, publish, and expound, such trifles of antiquity as a modern author would be contemned for."[46] A similar sentiment was echoed in the May 21, 1711, edition of *The Spectator*. Here the author contrasts a "Gothic manner of writing" that appeals to an artificial taste, with a style of writing "that […] pleases all kinds of palates." Homer, Virgil, or Milton, "so far as the language of their poems is understood, will please a reader of plain common sense, who would neither relish nor comprehend an epigram of Martial or a poem of Cowley." By the same token, only affectation or ignorance will pre-vent anyone from enjoying "an ordinary song or ballad, that is the delight of the common people."[47]

Here common sense is associated not only with the culture of commoners but also with the classical tradition, while pedantry is made synonymous with German letters. A month later, Addison writes that "The worst kind of pedants among learned men, are such as are naturally endued with a very small share of common sense, and have read a great number of books without taste or distinction." While learning "finishes good sense," it gives a silly man "an opportunity of abounding in absurdities."[48] In a similar vein, another issue of *The Spectator* argues that "music, architecture, and painting, as well as poetry and oratory, are to deduce their laws and rules from the general sense and taste of mankind, and not from the principles of those arts themselves."[49] This was a view that would be developed at length by the popular painter and engraver William Hogarth in *An Analysis of Beauty* (1753),[50] while the self-taught poet Alexander Pope wrote in 1709 that

> by false learning is good sense defaced:
> Some are bewilder'd in the maze of schools,
> And some made coxcombs nature meant but fools.
> In search of wit these lose their common sense,
> and then turn critics in their own defense.[51]

Although common sense was thus generally spoken about as a natural endowment, in *The Spectator* it was understood to be something different than instinct.[52] Yet common sense was not synonymous with common opinion or fashion, either. A passage from the December 27, 1711, edition of *The Spectator* makes this point, as the unidentified author tries to come to grips with the essential meaning of the term. In a discussion of rules of etiquette, the author suggests that one should not need them for some things, like "outward civilities and salutations":

> These one would imagine might be regulated by every man's common sense, without the help of an instructor; but that which we call common sense suffers under that word; for it sometimes implies no more than that faculty which is common to all men, but sometimes signifies right reason, and what all men should consent to. In this latter acceptation of the phrase, it is no great wonder people err so much against it, since it is not every one who is possessed of it, and there are fewer who, against common rules and fashions, dare obey its dictates.[53]

Earlier that year Steele had suggested that "instances might be given, in which a prevailing custom makes us act against the rules of nature, law, and common sense,"[54] again indicating that common sense was understood to have more to do with right reason than with common opinion or custom. It certainly had nothing to do with the follies arising from aristocratic codes of honor, or with the self-regard of women who fall prey to the flatteries of the "woman's man," or with "narrow party humour" that "destroys virtue and common sense, and renders us in a manner barbarians towards one another."[55]

* * *

The latter use of the term, denoting a sense or regard for society as a whole, points to the Earl of Shaftesbury's influential essay, first published in 1709, titled *Sensus Communis. An Essay on the Freedom of Wit and Humour*. Anthony Ashley Cooper, The Third Earl of Shaftesbury (1671–1713), had an impact on eighteenth-century thought all out of proportion to his brief lifespan. His *Characteristicks of Men, Manners, Opinions, Times* (1711), which included the essay *Sensus Communis*, went through 11 editions by 1790 and influenced thinkers throughout Europe.[56] Shaftesbury was a truly seminal figure in modern intellectual history who, as an opponent of the skeptical and materialistic strains of modern thought, foregrounded "common sense" as a term of philosophical and social importance.

Shaftesbury argued, against Hobbes and others who emphasized human selfishness and depravity, that human beings are naturally sociable and there is no essential conflict between public and private interest. Moral virtue consists in serving the common or public good; a moderate degree of selfishness is also "essential to goodness" to the degree that caring for oneself serves the public good. To serve the common good is to serve oneself, since human beings live and move and have their being in society. Besides being naturally disposed toward ourselves, we have natural social affections and native moral principles, and virtuous behavior is the result of recognizing and acting upon these affections and principles so as to achieve a harmonious ordering of public and private affections.[57] "Common sense" emerges in Shaftesbury's discourse as an intuitive apprehension and affection for the public good, a *"Sense* of the *Public Weal*, and of the *Common Interest*; Love of the *Community* or *Society*, natural Affection, Humanity, Obligingness, or that sort of *Civility* which rises from a just *Sense* of the *common Rights* of Mankind, and the *natural Equality* there is among those of the same Species."[58]

In the essay *Sensus Communis*, Shaftesbury styles himself a champion of friendly conversation and raillery, in opposition to the musings of solitary thinkers who propound theories at odds with the breadth of common experience. And, he says, it became clear after one such conversation that although everyone appealed to common sense, no one seemed to know what it was. According to one gentleman who was present, if one meant by the term common opinion or judgment, then "'That which was according to common Sense to day, wou'd be the contrary to morrow, or soon after.'"[59] There seemed to be no enduring *sense* of religion, morality, or politics that was shared in *common*. In a jibe at his former tutor John Locke, Shaftesbury notes that "even [some] of our most admir'd modern Philosophers had fairly told us, that *Virtue* and *Vice* had, after all, no other *Law* or *Measure*, than *mere Fashion* and *Vogue*."[60] Shaftesbury's aim in the essay, then, is to "try what certain Knowledg or Assurance of things may be recover'd" by questioning and ridiculing received opinions, just as modern skeptics had.[61] And such a quest is best conducted in an open, humorous, and dialogical manner, free from the "ridiculous Solemnity and sour Humour of our *Pedagogues*."[62] Reason has more to do with the freedom, wit, and humor of a gentleman's club than with solemn assemblies that keep understandings at a distance. "I can hardly imagine that in a pleasant way [men] shou'd ever be talk'd out of their Love for Society, or reason'd out of Humanity and *common Sense* […] Philosophical speculations, politely manag'd, can never surely render Mankind more unsociable or un-civiliz'd."[63] For Shaftesbury, "politeness" was an ideal of discourse aimed at harmonizing philosophical speculation with an expanding public sphere of commerce and sociability.[64]

Shaftesbury's *Essay* thus seeks to isolate and identify the enduring nature of *sensus communis* in a light-hearted and amusing manner, as opposed to the seriousness of humorless divines as well as "fierce Prosecutors of Superstition" like Thomas Hobbes, whose philosophy made himself and mankind appear "savage and unsociable."[65] Shaftesbury argues that there is in fact such a thing as goodness, that it consists in disinterested service to others, and that it is in accord with common sense: "He who wou'd frankly serve his Friend, or Country, at the expence of even his Life, might do it on fair terms […] 'Twas *Inviting* and *Becoming*. 'Twas *Good* and *Honest*. And that this is still a good Reason, and according to *Common Sense*, I will endeavour to satisfy you."[66]

Such public-spirited behavior is the result of a "*Sense of Partnership* with human Kind" that arises in egalitarian publics.[67] Since despotic states do not allow for the existence of such publics, virtue cannot thrive in them, and hence morality and good government go together.[68] Britons can be happy that they have a better sense of government than their ancestors: they now have a notion of a "Publick" and of balanced constitutional government, which yields maxims that "are as evident as those in *Mathematicks*. Our increasing knowledg shews us every day, more and more what COMMON SENSE is in Politicks. And this must of necessity lead us to understand a like *Sense* in Morals; which is the Foundation." Faith, honesty, justice, and virtue are examples of such common-sense moral principles that must have existed in a state of nature as preconditions for civil union.[69]

In advancing this view, Shaftesbury is typically seen to have inaugurated the "moral sense" philosophical tradition in Britain, and there can be little doubt that he was enormously influential in this regard. But, as we have seen, British opponents of skeptical ideas had long appealed to innate moral principles, even if they tended not to speak of a "natural *Sense* of Right and Wrong" as Shaftesbury did.[70] Even then, the Cambridge Platonist Henry More suggested, half a century before Shaftesbury, that everyone has a "natural *Sense* or *Remorse of Conscience*" that is an undeniable witness to our moral freedom.[71]

Human beings' common moral sense can be obscured by disordered passions and philosophical subtleties. The thought of doing a *little* villainy or committing *one* treachery

> is the most ridiculous Imagination in the world, and contrary to COMMON SENSE. For a common honest Man, whilst left to himself, and undisturb'd by Philosophy and subtle Reasonings about his Interest, gives no other Answer to the thought of Villainy, than that *he can't possibly find in his heart* to fret about it, or conquer the natural Aversion he has to it. And this is *natural* and *just.*[72]

Honesty has little to gain by deep speculations of any kind:

> In the main, 'tis best to stick to *Common Sense*, and go no further. Mens first Thoughts, in this matter, are generally better than their second: their natural Notions better than those refin'd by Study [...] According to common Speech, as well as common Sense, *Honesty is the best Policy*: But according to refin'd Sense, the only *well-advis'd* Persons, as to this World, are *errant Knaves.*[73]

Shaftesbury thus arrays plain, natural common sense against the subtle speculations of philosophers like Hobbes, who ascribe all motives to selfish desires and private interest. But if he speaks of common sense as an intuitive moral sense and an affection for the common good, Shaftesbury also refers to it in more intellectual terms, as a faculty of judgment that is synonymous with "*fundamental Reason*" and "*natural Knowledg.*"[74] As in his moral and aesthetic theory generally, Shaftesbury's notion of common sense unites heart and head, part and whole, self and society, into an all-encompassing vision of a coherently ordered world. Early in his career, Shaftesbury had edited a collection of sermons by Benjamin Whichcote, a founding figure of Cambridge Platonism, and his thought reflects the worldview of that school: the world is an organic whole pervaded by spirit, rather than a lifeless machine made up of discrete parts. Each part bears some relation to the other parts, and truth entails understanding how each part fits into a rationally ordered whole. Sense and reason go hand in hand, in this perspective.[75]

Ridicule emerges in this context as "*one* of those principal Lights or natural Mediums, by which Things are to be view'd, in order to a thorow Recognition."[76] Skeptical truths are partial truths and need to be tested in all lights, including the light of ridicule, according to Shaftesbury. The appeal to common sense against the apparent absurdity of skeptical ideas was—and would remain—a common tactic used by opponents to skepticism, including philosophers of common sense. What Shaftesbury does, in *Sensus Communis*, is to provide a conceptual framework for understanding the role played by ridicule and a sense of absurdity in the discovery of truth. Truth is not the province of solemn private philosophers with narrow conceptions of human nature but rather a function of the various perceptions and forms of discourse of the human community at large. A sense of absurdity has something to do with how things appear to a gathered community. Although Shaftesbury's immediate "community" was a gentlemanly one, his discussion of common sense pointed to a social conception of truth grounded in human nature and an egalitarian public sphere that was emerging in eighteenth-century Britain.[77]

A linkage between ridicule and "common sense [… or] first notions that all men equally have about the same things" was in fact being made at the same moment in time by Françoise Fénelon, Archbishop of Cambrai. In his *Demonstration of the Existence of God* (1712, 1718), Fénelon asserted that laughter and ridicule are appropriate responses to propositions that a child or an ignorant laborer would find silly, as, for example, the notion that a table would be able to walk to another room. Such propositions are ridiculous because they shock common sense, a sense possessed by all, which discovers the absurdity of a question as soon as it is considered and makes us laugh. Common sense can thus be understood as those ideas or general notions that cannot be contradicted or inquired into and by which we examine and decide all things, according to Fénelon.[78]

Shaftesbury was thus not alone in highlighting "common sense" as a term with social and philosophical meaning; and he did so by drawing upon ancient conceptions of *sensus communis*. At a critical point in *Sensus Communis*, Shaftesbury includes a long footnote that documents an intensive study of the term as used by Greek and Roman writers. A "battle of the books" between "ancients" and "moderns" had been raging in Britain for some time,[79] and Shaftesbury stakes out his position in this battle by grounding his appeal to common sense in ancient sources, including particularly the satires of Juvenal and Horace. "Rare is common sense in men of [noble] rank," a line from Juvenal's *Satires* is interpreted by Shaftesbury to refer to a "*Sense of Publick Good*" not shared by princes, nobles, and courtiers.[80] Emerging from the traditions of ancient Stoicism, *sensus communis* was used by Roman writers to denote a sense of the shared mores and manners of the community, an instinctive point of contact between each individual and the thoughts and needs of the larger society.[81]

* * *

Shaftesbury developed and refined this meaning, as did his Italian contemporary Giambattista Vico (1668–1744), who advanced a holistic understanding of common sense that shared affinities with Shaftesbury's use of the term. Vico, a founding figure in the philosophy of history and culture, is reputed to have visited Shaftesbury in Naples and may have been familiar with the *Characteristicks*. But Vico conceived of *sensus communis* less as a natural affection than as "judgment without reflection, shared by an entire class, an entire people, an entire nation, or the entire human race." As Vico goes on to state in his *New Science* (1744), "Uniform ideas originating among entire peoples unknown to each other must have a common ground of truth. This axiom is a great principle which establishes the common sense of the human race as the criterion taught to the nations by divine providence to define what is certain in the natural law of the peoples."[82]

According to Leon Pompa, Vico's common sense is a kind of "communal essence of mind" that is able, in the course of experience, to grasp the need to maintain certain practices—for example, belief in Providence, burial of the dead, legal marriage—that form the basis for human civilization.[83] In the first edition of *New Science* (1725), Vico states that these three practices are "common senses of mankind:" a "vulgar wisdom" that began in religion and law "and reached its perfection and completion in the sciences, disciplines and arts."[84] The recondite wisdom of philosophers should aid, support, and be led by this vulgar wisdom of nations, rather than distort and abandon it, as Plato did.[85] Thinkers who do not work within the parameters of human nature are "monastic, or solitary, philosophers" who rend the nature of man rather than raising and directing it.[86]

For Vico, to understand human culture and history one must understand the essential structure and beliefs of the human mind. As he put it in the first edition of New Science, since "the world of gentile nations was […] made by men," one thing is for certain:

> The principles of this world must be discovered within the nature of the human mind and through the force of our understanding, by means of a metaphysics of the human mind. Hence metaphysics […] must now be raised to contemplate the common sense of mankind as a certain human mind of the nations, in order to lead the mind to God as

eternal Providence, which would be the most universal practice in divine philosophy.

Such a metaphysics must not be hypothetical but based on fact, that is, on "the modifications of our human mind in the descendants of Cain before the Flood, and in those of Ham and Japhet after it."[87]

Although Vico scholars tend to focus attention on Vico's discussion of the variation in the forms of common sense over time,[88] Vico clearly conceives it to be both a universal feature of the human mind and the particular beliefs and judgments that are manifestations of this universal mind in any one age or culture. Cross-cultural variations of proverbs are for Vico one example of the way in which *sensus communis* produces particular manifestations of universal wisdom.[89] Vico's common sense is a profoundly social and historical entity that is nonetheless ultimately rooted in human nature and a providential natural order.[90]

* * *

Shaftesbury and Vico may thus be seen as having developed, each in their own manner, a holistic notion of common sense in response to the perceived narrowness of modern scientific thought: Hobbesian egoism in the case of Shaftesbury and Cartesian rationalism in the case of Vico.[91] Both thinkers saw knowledge and truth more as the products of discursive communities than of solitary individuals; "common sense" joins individual minds and hearts together into larger wholes that provide a kind of universality and certainty to knowledge claims that is unavailable to solitary thinkers.[92] As we have seen, such a viewpoint had been present in early responses to modern skepticism and the ideas of Locke. For their part, Shaftesbury and Vico explored the deeper meaning of such an appeal and began to foreground "common sense" as a term that expressed the affective and social dimensions of human knowledge. Subsequent philosophers of common sense, while employing the term more narrowly in reference to the constituent features of individual minds, would keep alive the notion that philosophical knowledge was in important respects a social product that must remain attuned to its roots in the truths and perceptions of common experience.

Chapter 6

COMMON SENSE AND MORAL SENSE: BUFFIER, HUTCHESON, AND BUTLER

The French Jesuit scholar Claude Buffier (1661–1737) was the first modern thinker to develop a fully fledged philosophical doctrine of common sense (*sens comun*). In his philosophical writings, Buffier sought to shore up connections between the human mind and the external world that had been left tenuous by Descartes and Locke. The inherent skepticism and idealism of modern philosophy did not square with the rough and ready perceptions of the average person, and Buffier sought to ground higher thought in common-sense experience.

Like most common-sense philosophers who followed after him, Buffier was an educator, and consequently he was very responsive to the growth in literacy and the rise of a modern public sphere of enlightened discourse that was in principle open to everyone. As a young man he taught courses in the humanities, and after joining the Jesuit order in 1695 he became a full-time man of letters who published popular works of history, philosophy, religion, geography, ethics, grammar, and logic, culminating in his 1732 *Cours de sciences*, a complete course of studies for youth. He was a sociable figure who moved in aristocratic circles and attended fashionable *salons*, yet he remained at heart a popularizer who aimed to make higher learning accessible to the average person. Buffier spearheaded the trend in Jesuit education toward studying history and geography as independent subjects, and his widely adopted textbook on French grammar helped to focus more attention on the vernacular. Buffier, who frequented the salon of the feminist Marquise de Lambert, was also a supporter of female education, arguing that women are as intelligent as men. He has thus been seen as "a radical and reforming element within the structure of Jesuit education."[1]

Buffier was a *scriptor* (scholar/writer) at the intellectually vibrant Collége Louis-le-Grand in Paris from 1701 until his death in 1737, and he was a founding editor of the *Mémoires de Trévoux*, a cosmopolitan journal that featured articles and reviews on a variety of learned topics, written mainly by Jesuits but also by other members of the European Republic of Letters, including Leibniz.[2] In 1724 Buffier published *Treatise on First Truths and the Source of Our Judgments*, followed in 1725 by a popularized account of his philosophical ideas entitled *Elements of Metaphysics Within the Reach of Everyone*. In both books Buffier presents his doctrine of *sens comun*. Thomas Reid was later accused of plagiarizing Buffier, but there are some differences between the ideas of each philosopher, and it is unlikely Reid was familiar with Buffier's work until after the publication of Reid's *An Inquiry into the Human Mind, on the Principles of Common Sense* in 1764.[3] Nevertheless, in responding to some of the same currents in early-modern thought, both philosophers shared fundamentally similar ideas and perspectives. Reid indicated his affinity for Buffier in 1785, stating that Buffier was "the first, as far as I know, after Aristotle, who has given the world a just treatise upon first principles."[4] Reid may have been a century off the mark (Herbert of Cherbury's *De Veritate* was published in 1624), but Buffier was the first modern philosopher to make "common sense," with its populist connotations, a central and defining feature of his philosophy.

Buffier openly acknowledged his debt to Descartes, Malebranche, and Locke and saw himself to be steering a middle way between the extremes of dogmatism and skepticism.[5] If on the one hand he argued for the existence of first truths of common sense that assure us of the existence of external objects and other minds (supplementing the Cartesian *cogito*), he also sought to differentiate between the propositions of which we can be certain and more relative human opinions.[6]

Similar to Descartes, Buffier locates the first source and principle of all truths in what he calls the "internal sentiment" [*le sentiment intime*] or feeling of our own existence: "I think, I feel, I exist."[7] This internal sentiment is seen by Buffier as the model for all truths, including those of common sense.[8] But if this internal sentiment of our own existence were the only "given," we would have to face the disastrous consequence of not being certain of the existence of physical bodies.[9] In order to avoid such extravagances, we must accept that there are other certitudes besides this first one that can be assigned to another head or rule of truth—the "common sentiment of nature or, as is ordinarily said, common sense [*sens comun*]."[10]

For Buffier, common sense is "the disposition that nature placed in all men, or manifestly in most men, in order to bring them to make common and uniform judgments—when they have reached the age of reason—on objects different from the internal sense of their own perception. Such judgments are by no means the consequence of any prior principle," rather they are the "incontestable and plausible principles of everything that a reasonable man is able to know."[11] Examples of such natural judgments include: there are other beings and men in the world; there is something in them called truth, wisdom, and prudence; there is something in me called mind or intellect, and something that is not which is called body; and all men are not in accord to mislead me. "These truths are such that if anyone were led to disagree we would be unable to dispense with viewing them as having lost their minds."[12]

First truths of common sense are therefore those truths which no sane person engaged in the practical affairs of life can seriously doubt, despite all the embarrassing philosophical subtleties and sophisms that can be presented in opposition to them.[13] There is thus a strong pragmatic element to Buffier's philosophy that would also appear in Reid's: the common experience of everyday life is a key arbiter of truth, and consequently any philosophy that calls into question the truths of common experience has something seriously wrong with it. Buffier's philosophy was hence aimed at overcoming the emerging modern dualism between higher thought and practical reason.[14]

Other first truths include the notions that man is free, that the existence of order implies intelligence, and that the mind or soul produces changes in the body.[15] The existence of God, however, is not a first truth, although there are first truths from which one can easily conclude the existence of God.[16] Metaphysical arguments that attempt to prove the existence of God from our own ideas, however, only demonstrate what is inherent in our ideas. If the idea of God necessarily includes existence, for example, all that shows is that we cannot form an idea of God that does not include his existence.[17]

Buffier, in responding to Locke, did not equate common sense with innate ideas. Ideas in Buffier's view are thoughts or representations of things, while by common sense he meant a disposition to think certain things on certain occasions. And if by "innate" we mean an idea or principle that is continuously perceived, it is obviously wrongheaded to admit innate first principles, since the same ideas are not always present in our minds. But if "innate idea" is used to signify a *disposition* to make certain judgments, Buffier says he will not quibble with the term. Locke's polemic against innate ideas that were (or should be) always present to the mind was superfluous, according to Buffier, since he was attacking a straw man.

Buffier also took issue with Locke on universally held moral principles. Locke, as we have seen, doubted if any existed, whereas Buffier suggested that the notions of fidelity and justice were universal moral principles, despite local variations on how justice and fidelity are conceived.[18] Buffier turns out to have been on solid ground—there is in fact good reason to believe that notions of justice (or reciprocity) and fidelity to promises are universal.[19] Despite such critiques, however, Buffier considered himself to be a follower of Locke.[20]

Besides asserting the existence of universal moral principles, Buffier argued that universal consent is a strong indicator of first truths. In his view, authority is a collective agreement—the witness of men gathered together is "a rule of truth that triumphs over my [individual] judgment."[21] The universal testimony of human beings is authoritative and conveys "moral certitude," and it is this moral certitude, which is less vivid but just as real as the basic certainties of sense perception, that guides us in many of our practical and social undertakings.[22] When one finds uniform sentiments, "it is human nature that is speaking," and only a false philosophy would contradict the universal sentiment of nature rather than working to clarify and understand it.[23] The false philosophy Buffier has in mind includes the skepticism of Bayle and ancient authors and the idealism of Berkeley.[24] Buffier goes so far as to provide, in the *Elements of Metaphysics*, stock answers that anyone can use to counter philosophical skeptics who might attack common sense.[25] Given that the author of nature can be supposed to have inscribed in each man the ability to know certain truths, "Every man, at least regarding some first principles, is as philosophical and credible as Plato and Descartes," and in some cases the average person is *more* credible than philosophers.[26]

It is clear from Buffier's ongoing polemic against philosophers that he feels modern philosophy needs to be brought back to its roots in common sense. "Philosophers who reject verbally (for they cannot do it in their heart of hearts) the principles and notions which are fundamental to our intelligence are intelligible after their own fashion. What they call metaphysics everyone else calls extravagance. True metaphysics doesn't seek to destroy the nature of things, but to consider it in its different lights."[27] A philosopher should not find it necessary to renounce common sense: "rather let us make common sense the basis of our philosophy."[28]

For Buffier, then, what passes for truth is a collective agreement grounded in human nature and the practical requirements of life, and it is furthermore a truth of common sense that those truths or facts that are affirmed by the experience and testimony of all human beings, in all ages, are true.[29] By the same token, if the "greater number" of human beings would not accept a proposition, it is proper to doubt that it is a first truth.[30] However, Buffier does not claim that truth necessarily resides in the opinions of the multitude.[31] As Louise Marcil-Lacoste suggests, "Buffier's persistent claim is that we must distinguish common sense from vulgar opinions and prejudices."[32] The reliable truths of the multitude are those that always carry along the greatest number of minds and are accessible to all who have reached the age of reason, while the unreliable ones are those truths that cannot be known without study, attention, and experience.[33] The multitudes *can* be relied upon when it comes to fundamental first truths, but truths of the "sciences" (broadly conceived) are another matter. "Taste" in literature, design, painting, and music involves external truths dependent upon specific knowledge, experience, and expertise.[34] But even though these kinds of truths may have something arbitrary about them, Buffier insists that nothing should prevent one from calling any judgment that nature causes men to make in common and in the greatest number "first truths," even on the most particular subjects. "As one is more sure about what is seen by the eyes of many than by one pair alone, one is also more sure about what is judged true by several minds than by one man."[35]

Knowledge is therefore ultimately a social, rather than individual, construct. When the vast majority of human beings' individual perceptions agree, one can be reasonably certain of the existence of a first principle, and this itself is a first truth.[36] By the same token, if human opinions vary such that universal opinion is truly ambiguous, then there are grounds to conclude that first principles of common sense do not exist on those topics. For example, given the relativity of views on human beauty across cultures, it may be impossible to find "an essential and real character of beauty."[37] The same goes for law: "If common sense were clearly in favor of one law, the great majority of humankind would have established it, instead of each nation or province remaining still attached to its own laws."[38] Truths of common sense are the substratum of human beliefs that remain constant beneath the variety of human opinions. Consideration of all points of view—something which Buffier values—highlights prejudices while at the same time helping to reveal the actual first truths of common sense.[39]

When it comes to natural science, Buffier feels that although we can be assured of the existence of external objects and of some of their basic qualities, we cannot know the intrinsic nature of things, and here his thought reflects the mitigated skepticism of modern science.[40] As a follower of Locke, Buffier accepted the basic mental framework in which "ideas" are images or representations of external bodies. While we may have a clear understanding of our ideas and their connections, they are abstractions that may or may not agree with external realities. Scientific axioms are internal logical systems or connections of ideas without any necessary hold over objects in the world.[41] Although our senses inform us of the existence of bodies, there is an infinity of dispositions of matter that the senses are too weak to report.[42] Because of this weakness of our senses, the "intimate physical constitution of matter is unknown to us [...] Its most perceptible qualities are impenetrability, mobility, and quantity."[43] The "secret springs" of nature are therefore hidden from us, and "everything that we know of matter and form [are] only vague ideas that allow us to know natural bodies only very superficially."[44] A pure system of science is akin to a "romance"—a conjectural history of facts that can be narrated differently until "confirmed by the senses united with the light of reason."[45]

As this latter statement indicates, even when he is at his most diffident about our ability to know the truth about something, Buffier still holds out the possibility for the attainment of some degree of certainty, if a common sense of the matter can be established. As long as the reports of our senses are not contradicted by reason, by an earlier or concurrent report from the same senses, or by the sense experience of other men, the evidence of those senses can be taken as "a type [*genre*] of first truth."[46] The senses themselves always give a faithful report of what appears to them: When our eye tells us that a ship is either advancing toward us or that the shore is receding, it is faithfully reporting appearances. It is up to the soul to discern the real state of things.[47] Buffier thereby defends the senses as being capable of producing a type of certainty, as far as they are capable of penetrating into the constitution of the world. In this respect Buffier was more of an empiricist than Descartes, who undermined the reliability of sense experience while asserting our ability rationally to grasp the essential features of external objects.[48]

* * *

Buffier was in his day quite influential in a variety of fields, and a number of his philosophical ideas—particularly those concerning sense experience— were incorporated into the *Encyclopédie* of Diderot and d'Alembert.[49] As an avowed follower of Locke, Buffier helped to disseminate Locke's thought in France,[50] and as a critical follower of Descartes, Buffier supplemented the *cogito*, which gave scant assurance of external truths, with the *sens comun*, which lent authority to the senses and provided certain—if limited—knowledge of external truths.[51] In steering a middle course between skepticism and dogma- tism or prejudice, Buffier sought to establish those things that we, as individual human beings living in community, *can* know with certainty, while clarifying those opinions (or *types* of opinions) that are probable and/or the result of communal prejudices. He opened the door to a science grounded in common sense, but he sided with the general thrust of early-modern philosophy of science when it came to our inability to grasp with complete certainty the underlying nature of the external world. Although he did exercise an influ- ence on Spanish Catholic thinkers,[52] Buffier's *Traité* seems to have been largely forgotten by the early nineteenth century, when Reid's thought was coming into vogue in France. Buffier's philosophy was nevertheless an important early philosophy of common sense, as he sought to locate the foundations of human knowledge and "truths," including truths about the external world, in the common intuitive perceptions of all human beings.[53]

The profoundly social and intersubjective orientation of Buffier's philosophy was clearly in step with early-modern social and intellectual developments, including the growth in literacy and the rise of print culture, the expansion of education, and the emergence of a relatively egalitarian realm of civil society that stood apart from entrenched political structures and hierarchies. A "public" with its own "opinion" was starting to emerge in eighteenth-century France (as elsewhere), as private individuals, guided by aristocratic ideals of civility and sociability, began to envision a realm of "society" that stood apart from the Crown.[54] Just as English writers had already begun to do, Buffier employed the term "common sense" in such a way as to link the spontaneous and "natural" perceptions of the average human being with higher thought and put a check on what were felt to be speculative—and absurd—learned excesses. The touchstone for "truth" was no longer the isolated, abstracted philosopher working alone in his study, it was now society at large.

Buffier himself was a profoundly sociable person. He was a popular figure who delighted in frequenting *salons* and dinner parties, he was one of a number of coeditors of the *Mémoires de Trévoux*, and the bulk of his literary output was aimed at making higher thought accessible to the widest possible readership. What is more, Buffier followed up his philosophical treatises of 1724 and 1725 with a *Treatise on Civil Society*, in 1726. In this book, Buffier extends the social orientation of his metaphysics to ethics, arguing that virtue is essentially social—good and bad are defined in terms of the benefit or harm one does to civil society, rather than according to Scripture or religious doctrine. Reflecting a growing trend, Buffier separates ethics from religion and joins them to manners, since goodness comes from pleasing others. Morality is defined by Buffier as "the science of living with other men in civil society in order to procure, as much as we can, our own happiness in concert with the happiness of others."[55] Human beings are able spontaneously to feel compassion and goodwill—*bienfaisance*—toward others, a view which lends a certain dignity and worth to the passions, when properly governed by reason. In his *Treatise on First Truths*, Buffier often refers to first truths of common sense as sentiments of nature, so it follows that his ethics would be built around "natural" moral sentiments.[56] All of this was enough to make Buffier an important figure in the development of French civil society, a "citizen without sovereignty" who began to articulate an ethos of civility that was starting to emerge within an absolute monarchy.[57]

* * *

If the *Tatler* and *Spectator* had begun to express such an ethos in early-eighteenth-century London (see Chapter 5), *Common Sense: Or, The Englishman's Journal* (1737–44) continued to develop the notion that "common sense" was a fundamental quality shared by members of civilized—and indeed *English*—society. The "plain good/common sense" of the average Englishman had by the 1730s become a common rhetorical figure, enough so that the journal, edited by Charles Molloy (d. 1767), could use the expression to signify any attempt to move beyond political partisanship to a more universal plane of civil discourse. Although it was essentially an opposition paper to the Walpole administration, *Common Sense* sought, especially in its first two years of publication, to achieve a more worldly, sociable, and witty tenor than other such papers. It was widely read and distributed and even had short-lived imitators in *Country Common Sense* and *Old Common Sense*.[58]

The immediate inspiration for the title of the periodical was a hit play of the 1736 theater season by Henry Fielding entitled *Pasquin* [...] *a Tragedy call'd, The Life and Death of Common-Sense*. In the play, a doctor, a lawyer, and a priest all express their antipathy to "Queen Common Sense," who abridges the power of each. While she is on earth, "Lawyers cannot rob men of their Rights; Physicians cannot dose away their Souls: [and] A Courtier's promise will not be believ'd." Firebrand, the priest, accuses Queen Common Sense of being an enemy of the Sun (his god), to which she replies that "I will never adore a Priest, who wears Pride's Face beneath Religion's Mask, and makes a Pick-Lock of his Piety, to Steal away the Liberty of Mankind." This is enough to cause Firebrand to murder Queen Common Sense, who meanwhile has joined battle with "Queen Ignorance" and her army of French and Italian singers, fiddler, tumblers, and rope dancers. Queen Ignorance claims to have been asked to come on behalf of Queen Common Sense's subjects, who "say you do impose a Tax of Thought Upon their minds, which they're too weak to bear." After her death, Queen Common Sense reappears as a ghost who scares off Queen Ignorance and counsels that "all henceforth, who murder common sense, learn from these scenes that tho' success you boast, you shall at least be haunted by her ghost."[59] Common sense is thus presented by Fielding as a levelheaded, critical check on the power of educated elites—particularly on their ability to lord it over the less educated by self-serving jargon—while also being identified as a particularly "English" cultural trait in the battle against foreign entertainments.

Although Fielding was an occasional contributor, *Common Sense: Or the Englishman's Journal* employed the term in a broader way, more in keeping with its use in the *Tatler* and the *Spectator*, and the journal in fact identified itself with those earlier weeklies.[60] The opening editorial, believed to have been written by Philip Stanhope, Fourth Earl of Chesterfield (1694–1773), set the tone for the journal by stating that the design of the paper was to "take in all Subjects whatsoever, and try them by the Standard of Common Sense," which the author believes to be "call'd common, because it should be so, [rather] than because it is so; it is rather that Rule by which Men judge of other Peoples Actions, than direct their own; the plain Result of right Reason admitted by all, and practised by few." The author goes on to identify "the old solid

English Standard of Common Sense" with the English constitution and suggest that common sense is a court with "numerous branches," such that "a King and a Cobler without [common sense] will equally bungle in their respective callings." The overall intention of the journal was "to rebuke Vice, correct Errors, reform Abuses, and shame Folly and Prejudice, without Regard to any Thing but Common Sense."[61]

In subsequent numbers of the journal, "Common sense" is identified in Shaftesburyean fashion with public spirit and the larger interests of the community,[62] as "a kind of negative wisdom" which points out "excrescences of affectation, fashion, party, and passion,"[63] as a sense of moral universals,[64] and as the arbiter of proper dress according to age, sex, figure, and social station.[65] Extravagances and immoderate behavior of all kinds are censured by the journal, since "Mankind have a strange inclination to branch out into Extreams, and will be dilating themselves into the Ridiculous, unless some judicious hand takes the trouble to prune their Luxuriances, and by that means make them bear the fruits of Common Sense." One must, in other words, "polish away [the] Crusts and Excrescences" of "Epidemical Frenzies" in order to "bring the World to perceive the Lustre of Common Sense."[66]

* * *

Given such fulsome praise in a weekly journal with "common sense" on the masthead, there can be little doubt that by the 1730s the term had truly "arrived" in Britain as a concept bearing intellectual weight and social significance. According to Sophia Rosenfeld, "Common sense came to provide a solid, if modest, epistemic and stylistic foundation for a wide variety of disciplines, including law, philosophy, history, natural science, and literature" that promised to prevent excessive factionalism, sectarianism, and social strife.[67] While the rise of "common sense" thus had many dimensions, our focus in this study is the way in which the idea and philosophy of common sense served to bridge the widening chasm between modern science and philosophy, on the one hand, and the everyday sense experience and perceptions of the rising "commons," on the other. This development became explicit in the thought of George Turnbull and Henry Home, Lord Kames, and reached full fruition in the work of Thomas Reid, who was Turnbull's student at Marischal College in Aberdeen, and Kames's friend. But before considering the thought of Turnbull, Kames, and Reid, we need to examine the development of Shaftesbury's moral intuitionism in the works of Francis Hutcheson and Joseph Butler, as an important feature of Reid's intellectual context. Reid himself, late in life, acknowledged that he was in full agreement with the moral sense doctrine of Shaftesbury and Hutcheson, which he saw to be congruent with his own doctrine of common sense; and while in his twenties he had taken extensive notes on Butler's *Analogy of Religion*.[68] Hutcheson's naturalistic and empirical approach to moral philosophy set the stage for an increasingly inductive and introspective "science of the human mind" that came to fruition in the work of Reid and the thinker who stimulated him more than any other, David Hume.

Hutcheson, who is generally acknowledged to be a founding figure in the Scottish Enlightenment,[69] was Professor of Moral Philosophy at the University of Glasgow from 1730 until his death in 1746. He was a popular and frequent lecturer who counted Adam Smith among his students, and his work was widely read and circulated in Europe as well as the American colonies. Born in Ireland, Hutcheson entered the University of Glasgow in 1710; in 1718 he returned to Ireland where he was licensed as a Presbyterian Minister. He ran a dissenting academy in Dublin for ten years, and it was during this period that he contributed articles to *The London Journal* and *The Dublin Weekly Journal* that established his lifelong concern to oppose the "selfish" philosophy of Thomas Hobbes and especially Bernard Mandeville, whose *Fable of the Bees: Or, Private Vices, Public Benefits* was first published in 1714. If Locke's attack on innate ideas in his *Essay* had stimulated further reflection on moral and epistemological first principles, Mandeville's stress on self-love as a powerful civilizing agent prodded Hutcheson and Butler to refine and develop Shaftesbury's notion of a "moral sense" as an essential feature of the human constitution and, by extension, social interaction.[70] In doing so, they each (in their own manner) helped to focus attention on natural faculties or dispositions—internal "senses" analogous to the external senses—that not only condition our experience of the world and make possible various forms of belief and knowledge but also lie at the very foundations of human sociality.

If Thomas Hobbes was the great bugbear of later seventeenth-century moralists, Bernard Mandeville played a similar role in the eighteenth century, due to his brilliant, rhetorically persuasive anatomization of man as a selfish animal who becomes civilized by the skillful harnessing and management of his passions. Enlarging upon an earlier poem titled *The Grumbling Hive: Or, Knaves Turn'd Honest* (1705), *The Fable of the Bees* presented a compelling—and "scientific"—account of human nature and sociability that was both troubling and stimulating for many eighteenth-century thinkers.[71] In Mandeville's view, an empirical analysis of human nature reveals that pride, self-regard, and other selfish passions are the foundations of all moral, public-spirited behavior and that without such passions, large, commercial polities would not flourish. Luxury and avarice help to keep the wheels of trade and manufactures turning, as do profligacy and even robbery, which can make inert wealth productive by spreading it around. Human pride is turned to useful ends by its ability to make human beings engage in behaviors—like self-sacrifice—that they would otherwise find repugnant, since those who perform such acts are esteemed and praised. "Virtue" is thus nothing more than a name for those performances "by which Man, contrary to the impulse of Nature, should endeavour the Benefit of others." Behavior that only gratifies one's own appetites, on the other hand, is called "vice." Shame works to keep people from overindulging in such behaviors, making it an essential ingredient in politeness and sociability.[72]

If all of this sounds quite foreign to the Shaftesburyean enterprise, Mandeville himself suggested that "two Systems cannot be more opposite than [Shaftesbury's] and mine."[73] In his essay "A Search into the Nature of Society," first published in the 1723 edition of the *Fable*, Mandeville launched a direct and very personal attack on Shaftesbury and his views on human nature and morality. In the essay, Mandeville painted Shaftesbury as an indolent moralist who found it easy to allow his passions to lie dormant and to pretend to an imaginary notion of virtue free from self-sacrifice. But the truth was, a man such as Shaftesbury "may form fine Notions of the Social Virtues, and the Contempt of Death, write well of them in his Closet, and talk Eloquently of them in Company, but you shall never catch him fighting for his Country, or labouring to retrieve any National Losses."[74] Shaftesbury had preached public spirit but had not himself exhibited much public-spirited behavior. Virtue consists in action, and someone with Shaftesbury's advantages, if he were truly virtuous, would have exerted himself more for the betterment of his fellows. The "calm Virtues" of the *Characteristicks* "are good for nothing but to breed Drones, and might qualify a Man for the stupid Enjoyments of a Monastick Life [...] but they would never [...] stir him up to great Achievements and perilous Undertakings."[75] Rather, vices like pride, envy, and vanity are what bring men to exert themselves for the public good, and Mandeville draws on a wealth of social and introspective data showing how this could be so. Shaftesbury's notions, while high-minded, were "inconsistent with our daily Experience."[76]

Besides attacking Shaftesbury as a hypocrite, a good portion of "A Search into the Nature of Society" is devoted to reviewing common examples of moral and cultural diversity. Whether one looked at painting, fashion, burial customs, marriage, or religion, all featured irreducible variation across cultures, making it hard to find universal moral truths.[77] And Mandeville pours scorn on man's vaunted sociability and the critique of solitude that had become a staple of anti-skeptics:

Would not a Man be by himself a Month, and go to Bed before seven o'Clock, rather than mix with Fox-hunters, who having all Day long tried in vain to break their Necks, join at Night in a second attempt upon their Lives by Drinking, and to express their Mirth, are louder in senseless Sounds within Doors, than their barking and less troublesome Companions are only without?[78]

As Mandeville sees it, "Men of Sense" who can think for themselves will prefer their "Closet or Garden" to the society of all those shallow worthless souls who cannot bear to be alone.[79] Even if we grant an inherent love of company, "does not Man love Company, as he does everything else, for his own sake?" Friendships do not last if they are not reciprocal, and the most popular people of all are the good-humored folks who allow others to shine.[80]

Mandeville's *Fable* thus posed a direct challenge to Augustan self-understandings by arguing that human beings are essentially egoistic animals, and moral values are social conventions that harness the passions in such a way as to benefit the wider community. Mandeville thus presented, according to E. J. Hundert, "a world that frighteningly demanded from its members the relegation of their civic ideals to the realm of nostalgia."[81] Mandeville's science of man threatened to make a shamble of the anti-skeptical moral realism put forth by Shaftesbury and other English innatists, and Francis Hutcheson and Joseph Butler soon took up the task of defending the Shaftesburyean legacy.

* * *

Francis Hutcheson, of whom it was said that he could not give a lecture at the University of Glasgow without denouncing Mandeville, entitled his first book *An Inquiry into the Original of Our Ideas of Beauty and Virtue [...] in which the Principles of the Late Earl of Shaftesbury Are Explain'd and Defended, against the Author of the Fable of the Bees* (1725). In this and other texts, including notably his *Essay on the Nature and Conduct of the Passions and Affections, with Illustrations on the Moral Sense* (1728), Hutcheson took up Mandeville's challenge by formulating his own inductive account of human nature that steered a middle way between rationalistic and egoistic moral theories. Hutcheson argued that although "Ingenious speculative Men, in their straining to support an *Hypothesis*, may contrive a thousand *subtle selfish Motives*, which a kind generous Heart never dreamed of," unbiased attention to what passes in our own hearts reveals that human beings come equipped, in addition to self-love, with "kind" or "benevolent" affections as well as a "moral sense" that sanctions them. Hutcheson defined the moral sense as an internal sense or faculty that is analogous to our external senses, by which we recognize and approve of kind, unselfish actions and disapprove of unkind, selfish affections and actions. Our moral sense is hence an innate disposition or "Determination of our Mind" that approves or disapproves of perceived affections or actions spontaneously and without regard to either reason or volition.[82] It is universal and generally uniform among men and is unchanging like the law of gravitation or any other law of nature.[83] Reason can help calculate the moral effect of any one action, but by the same token it can be responsible for immoral behavior by making a false calculation of the public interest. This is one of the reasons for moral diversity.[84]

Virtuous actions are therefore the result of kind or benevolent affections for others and the public good and are justified by our moral sense; and both our kind affections and our moral sense are instinctive features of our constitution that afford us, when they are gratified, "our most intense and durable *Pleasures*."[85] This is a thoroughly naturalistic account of virtue,[86] and Hutcheson was in fact accused by church authorities of teaching, contrary to the Westminster confession, that it is possible to have a knowledge of good and evil without knowledge of God. Hutcheson had been pushed to such a potentially heterodox position by his attempt to provide an empirical and naturalistic answer to the "selfish" theories not only of Hobbes and Mandeville but also of the German thinker Samuel Pufendorf, who had adopted a Hobbesian view of human nature in formulating a natural law of human sociality that was free from theological presuppositions.[87] Pufendorf argued that since human beings are inherently selfish, they come to recognize—as a self-evident truth, no less—the fundamental law of nature, which is that "man ought to do as much as he can to cultivate and preserve sociality." Human self-interest dictates that we subordinate ourselves to society, or risk being destroyed by each other.[88] The upshot of all such theories, in Hutcheson's view, was "that the old notions of natural affections, and kind instincts, the sensus communis, the decorum, and honestum, are almost banished out of our books of morals; we must never hear of them in any of our lectures for fear of innate ideas: all must be interest, and some selfish view."[89]

Clearly, the only way to answer modern attacks on such "old notions" was by looking deep into human nature for them—any other response would be deemed unscientific.[90] Not only worldly gentlemen like Shaftesbury but Christian educators like Hutcheson would also have to take an empirical look at the human mental/psychological constitution as the basis for their moral arguments. As he stated to his University of Glasgow peers in 1730, rather than looking at men as being either embryonic blank slates or vicious brutes, "It would certainly be much more useful to inquire into [the] natural judgements, perceptions, and appearances of things that nature presents [...] For there are indeed many powers natural to all kinds of things, many senses and appetites in animals, many natural structures that are not at all apparent from the outset."[91]

Here was a research agenda for others to follow, put forth by the holder of the Glasgow Chair of Moral Philosophy; and many did, including notably Adam Smith, David Hume, and Thomas Reid, each in their own (often conflicting) ways. And there can be little doubt that Hutcheson's own account of common "internal" senses—including, in addition to the moral sense, a sense of beauty, a sense of honor, and a sense of public good—helped to direct Scottish philosophical attention toward naturalistic analysis of the human constitution as the locus of senses, dispositions, and instincts that govern our moral as well as intellectual perceptions, including those of the sciences.[92]

* * *

Joseph Butler (1692–1752), dean of St. Paul's Cathedral and eventually bishop of Durham, also pushed philosophical analysis in a naturalistic direction, concerned as he was to refute skepticism and defend orthodoxy by recourse to the natural constitution of the world and human nature. In his oft-reprinted *The Analogy of Religion, Natural and Revealed, to the Constitution and Course of Nature* (1736), Butler sought to convince those who might be susceptible to skepticism that nature itself provides plenty of evidence for the truths of the Christian religion and that human beings are equipped with a conscience or moral sense that is well-suited to discern the moral order of the universe.[93]

The fundamental argument of *The Analogy of Religion* is that natural processes reinforce and illuminate the truths of revealed religion. For example, just as living beings undergo various transformations in this life (a mature adult is vastly different from a fetus in the womb), there is reason to believe that such transformations continue into the next life.[94] Sleep shows us that living powers continue to exist even when not exercised; there is consequently a good likelihood that living powers (i.e., the soul) continue after the death of the body.[95] There are daily instances of mercy and compassion in the general conduct of nature, as when human beings care for each other. "By a method of goodness analogous to this, when the world lay in wickedness, and consequently in ruin, 'God so loved the world, that he gave his only begotten Son' to save it."[96]

In making such analogical arguments, Butler eschews abstract speculation in favor of intuition and common sense. Because Christianity "carries in it a good degree of evidence for its truth, upon its being barely proposed to our thoughts," there is no need for "abstruse reasonings and distinctions, to convince an unprejudiced understanding, that there is a God who made and governs the world, and who will judge it in righteousness."[97] The "moral system of nature, or natural religion, which Christianity lays before us, approves itself, almost intuitively, to a reasonable mind, upon seeing it proposed."[98] What is more, belief in God as creator and moral governor of the world has been professed in all ages and countries; such "general consent [...] shows this system [of belief] to be conformable to the common sense of mankind," and hence must be given great weight.[99]

Whereas speculation leaves us uncertain as to whether good actions are linked to happiness, "the whole sense of things that [God] hath given us, plainly leads us, at once and without any elaborate inquiries, to think that it [...] must be to good actions chiefly that he hath annexed happiness." It is a plain and obvious fact that the intention of nature is to be a school of discipline for improving our character. Such matters of fact "ought, in all common sense [...] awaken mankind, to induce them to consider in earnest their condition, and what they have to do."[100]

Such appeals to intuition and common sense are repeated throughout Butler's *Analogy*. And in his appended essay "Of Personal Identity," which was of great interest to Reid, Butler suggests reasons why common sense and intuition are important elements of philosophical discourse. In the essay, Butler complains that some followers of Locke, who had taken his "hasty observations" on personal identity to an extreme, argue "that personality is not a permanent, but a transient thing" and hence that personal responsibility for our past actions is a fallacy.[101] On Butler's view, "The bare unfolding [of] this notion, and laying it thus naked and open, seems the best confutation of it [...] All imagination of a daily change of that living agent which each man calls himself, for another, or of any such change throughout our whole present life, is entirely borne down by our natural sense of things." Everyone, in other words, is conscious "that he is now the same person or self he was, as far back as his remembrance reaches."[102] But why trust memory?

> He who can doubt, whether perception by memory can in this case be depended upon, may doubt also, whether perception by deduction and reasoning, which also include memory, or indeed, whether intuitive perception can. Here we can go no further. For it is ridiculous to attempt to prove the truth of those perceptions, whose truth we can no otherwise prove, than by other perceptions of exactly the same kind with them, and which there is just the same ground to suspect; or to attempt to prove the truth of our faculties, which can no otherwise be proved, than by the use of means of those very suspected faculties themselves.[103]

Here Butler makes arguments that came to be fundamental to the school of common-sense philosophy. Although we can reason ourselves into all sorts of beliefs, and out of others, reason is but one faculty among others, and it must stay within its proper sphere. And there are certain truths that cannot be proven, because the very act of doing so presupposes what is to be proven.

In the case of personal identity, we cannot prove the existence of a continuous personal identity—or for that matter disprove it—because in the very act of doing so we assume a continuous self that is making the inquiry. And there is a pragmatic element to take into account as well: as Butler says, even if we could go against common sense and convince ourselves that we were not the same person as we were yesterday, we would still be unable to alter our conduct accordingly.[104]

In the end, for Butler

> The proper motives to religion, are the proper proofs of it, from our moral nature, from the presages of conscience, and our natural apprehension of God, under the character of a righteous Governor and Judge; a nature, and conscience, and apprehension given us by him; and from the confirmation of the dictates of reason, by *life and immortality brought to light by the Gospel.*[105]

Our God-given, natural human constitution not only gives us reliable knowledge about the world but is also capable of discerning the existence and truths of God and morality as they are revealed in both nature and Scripture. George Turnbull, Reid's teacher at Marischal College in the 1720s, would go on to develop this viewpoint, combining the natural law tradition stretching back to the ancients with a fervent enthusiasm for the discoveries and methods of Newtonian science, thereby laying the groundwork for a philosophy of common sense that could leave theological questions to one side and focus intently on a scientific analysis of the features of the human mind itself. Turnbull's synthesis and Kames's foregrounding of "common sense" as the answer to Humean skepticism were important antecedents of Reid's *rapprochement* of common sense and science, and it is to these seminal thinkers that we now turn.

Chapter 7

COMMON SENSE AND THE SCIENCE OF MAN IN ENLIGHTENMENT SCOTLAND: TURNBULL AND KAMES

Previous chapters have shown how a variety of European thinkers—including Herbert of Cherbury, René Descartes, Henry Lee, G. W. Leibniz, and Henry More, among others—were beginning to explore and develop the idea that human beings intuitively perceive certain notions, ideas, truths, or principles that condition our experience and make moral, scientific, and religious knowledge possible. We have seen how Claude Buffier was developing a philosophy of *sens comun* in eighteenth-century France and how the seeds of a philosophy of common sense were being planted in the British Isles by moral philosophers like Francis Hutcheson in Ireland and Scotland, and the Third Earl of Shaftesbury and Bishop Butler in England, where the figure of "common sense" as a term of popular culture was emerging in its own right.

Now we turn our attention to Enlightenment Scotland, where a fully-fledged philosophy of common sense, with strong links to modern science and the scientific method, began to emerge: first in the works of George Turnbull and Henry Home, Lord Kames, before coming to full fruition in the work of Thomas Reid. Although previous European thinkers had been making similar arguments and observations, it was in Scotland that a coherent and developed "school" of common-sense philosophy grew and matured, extending deep into the nineteenth century and far beyond the shores of Scotland.[1]

Hutcheson, Turnbull, Kames, and Reid were a subset of thinkers within the larger Scottish Enlightenment, which has come to assume a place in modern intellectual history that rivals the French Enlightenment. According to Donald Withrington, "The really distinctive mark of the Enlightenment in Scotland is that its ideas and ideals were very widely diffused, in all areas and among a very wide span of social groups, in what was for the time a remarkably well-educated and highly-literate population."[2] Alexander Broadie argues that the Scottish Enlightenment had roots stretching back to the fifteenth century and notes that its proponents were highly literate, sociable, and engaged in various intellectual clubs and societies, and as such "lived in each other's intellectual pockets."[3] Scottish thinkers were cosmopolitan members of the wider European Republic of Letters; according to Roger Emerson, "philosophical, medical and scientific ideas came from France and Holland as readily as from England; toleration and liberal theology were Dutch and Swiss as well as English; polite standards owed as much to the French as to Addison."[4] At the same time, it must not be forgotten that religious belief permeated Scottish life and intellectual inquiry and that religion was in many respects socially positive and creative.[5] According to Nicholas Phillipson, "the fear of scepticism and the desire to found a Science of Man that would serve the interests of Christians as well as unbelievers were among the hallmarks of Scottish learning in the age of the Enlightenment."[6]

Modern science and scientific thinking were especially important features of the Enlightenment in Scotland. According to Anand Chitnis,

> The pursuit of natural philosophy and of medicine [were] essential elements in the philosophic movement that is characteristic of eighteenth-century Scotland. Colin McLaurin, William Cullen, Joseph Black, James Hutton and Alexander Munro are as significant to the Scottish intellectual enterprise as Francis Hutcheson, David Hume, Adam Smith, Adam Ferguson and John Millar.[7]

Paul Wood has done much to flesh out the contours of Scottish science and the Science of Man that emerged in the Scottish intellectual context, arguing that "natural knowledge [was] a pivotal component of the intellectual culture of the Scottish Enlightenment" and that it "was used as a cultural resource in the ongoing religious disputes of the period, [occasioning] periodic anxieties about atheism and irreligion which focused public attention on the broader meanings of the scientific enterprise."[8]

*   *   *

The blending of religious and scientific perspectives is nowhere more apparent than in the writings of George Turnbull (1698–1748), a Scottish educator and moralist. Turnbull was an early and forceful advocate for the application of the methods and perspectives of the New Science—of Isaac Newton in particular—to the study of human nature and morality. If a gulf had opened up between common sense and scientific thinking in the modern age, Turnbull laid the groundwork for a *rapprochement* between the two that would find full expression in the thought of his pupil, Thomas Reid. For Turnbull, natural science "hath in itself a wonderful influence towards harmonizing the mind, and improving our taste and love of order, proportion, and harmony, and was for this reason called by Plato a purifier of the mind."[9]

As we have seen, the moral sense school emerged in response to the more skeptical and empiricist currents of thought that had arisen in tandem with the New Science. Turnbull was a contemporary of Francis Hutcheson and claimed, in his mature work, to have been influenced by him, along with Shaftesbury, Butler, and Pope, among others.[10] But Turnbull's graduation theses at Marischal College in Aberdeen, where he served as Reid's teacher in all subject areas, reveal that even before the publication of Hutcheson's earliest works, Turnbull's thought was already oriented toward forging a close bond between natural and moral philosophy that would provide a foundation not only for "scientific" exploration of the moral sense but for all the senses and faculties of the human mind—what Turnbull, following Henry More, referred to as the "furniture" of the mind.[11] In so doing, he harmonized the methods and findings of modern science with the emerging moral sense school, the natural law tradition stretching back to ancient writers, and orthodox Christian theology. The result was an all-embracing vision of a harmonious, orderly, and rational world in which all inhabitants are equipped by God with the mental faculties and tools to "make sense" of and flourish in that world as intended by their Creator, if such tools and faculties are properly understood and employed.

There are thus a number of components to Turnbull's philosophical system.[12] The place to begin is the Graduation Theses presented by Turnbull to students and faculty of Marischal College in 1723 and 1726 (Reid was noted to be in attendance at the latter event). The 1723 Philosophical Theses claimed that science is now on a "firm foundation" since "it is sustained not by fanciful hypotheses or unfounded conjectures, but entirely by either mathematical reasoning or clear and certain experiments and analogy." According to Turnbull, Newton's method of analysis and synthesis, so successful in providing mathematically demonstrable propositions about the material world, should—as Newton himself suggested—now be employed to extend the boundaries of moral philosophy.[13] Reasoning based on experiments, evidence, and analogy is most useful and convincing, "so perfectly are our minds adjusted to the condition of human life." By revealing order in nature, modern science disproves atheism by pointing toward a "uniform and intelligent cause" distinct from matter. And it has been confirmed by the whole science of matter and motion that "physical causes are simply natural laws and forces, fashioned and preserved with consummate skill by the Most Wise Creator of nature."[14]

Although great progress has been made in the scientific understanding of the natural world, understanding of the moral world is more limited. But "whatever the difficulty in ascertaining the administration of the moral world by the light of nature, one has to admit that the order of the corporeal world is most elegant and exceedingly neat. And it certainly offers us a very fine pattern of life and morals." This is because all its parts are united and move together in harmony, "banded together in a communion which is very well suited to preserve it and keep it safe." Thus one can argue that it is most pleasing to God, the Creator and Ruler of nature, "that all whom he has made to share in reason and a common sense (*sensus communis*) should pursue the common good of all intelligent beings."[15]

Here it is evident how Turnbull's philosophical approach, blending scientism with natural theology, promises a reconciliation between common sense and science, while also providing an agenda for further analysis of the mental faculties and powers that allow us to make sense of the natural and moral features of the world.[16]

Similar themes are to be found in Turnbull's Academical Theses of 1726, but he goes further in stipulating that the moral and physical worlds are systems within systems and that all parts of the world are mutually adjusted to each other. "And if we examine the powers and faculties of men and the material available to each individual, as well as the laws by which they can and should be directed, we shall quickly recognize that the human race is no less a regular system than the wonderful assemblage of the planets." The "special quality" of man is "the ability to learn the nature and causes of things [...] And nature supplies material to the senses with unstinting hand." Nature has bestowed upon man sufficient powers and natures "for his own good and the preservation of society" and "from these powers and laws of our nature reason may be rendered sufficiently well-adjusted [...] to the whole condition of mortal men." Thus it is quite clear that "the physiology which lays out the true order of the Constitution of the natural world must underpin moral philosophy. For such a Physics is nothing other than acquaintance with the mind which most perfectly rules all things."[17]

Turnbull would go on to develop these themes in his two-volume work, *The Principles of Moral and Christian Philosophy*, published in 1740 but drawing on his lectures while at Marischal College. Turnbull left the College in 1727, spending much of the rest of his life in peripatetic fashion, frequently serving as a tutor to wealthy families and traveling throughout Europe. He received a Bachelor of Civil Law degree from Oxford in 1733 and was ordained by the Church of England in 1739, eventually finding a post as rector of Drumachose parish in Ireland in 1742, which he held until his death in 1748. By the time he published *The Principles of Moral and Christian Philosophy*, he was a worldly and well-traveled citizen of the Republic of Letters, and the text exhibits his mastery of a wide range of ideas and texts, both ancient and modern.[18]

Volume 1 of the work, titled *The Principles of Moral Philosophy*, opens with the same themes discussed above: the aim of the treatise, in its most general aspect, is to account for moral phenomena using the methods employed by Newton to account for natural phenomena. Despite Scholastic abuses, moral philosophy must be considered a question of "fact or natural history, in which hypotheses assumed at random, and by caprice, or not sufficiently confirmed by experience, are never to be built upon." The study of mind must proceed along the same lines as the study of the human body, or any other branch of Natural Philosophy, with the goal of explaining moral phenomena "as the best of Philosophers teaches us to explain natural phenomena."[19]

As we have seen, this kind of methodological scientism was now considered the way forward by moralists interested in answering skeptics with a credible modern moral and religious philosophy. For many Scottish thinkers, including Turnbull and later Reid, Isaac Newton served as a kind of authoritative "anchor" for modern philosophy in the wake of the revolution in science and the often corrosive thinking that had emerged in its wake. Using the empirical and mathematical methods of the New Science, Newton had shown that the world is an orderly, law-abiding mechanism in which all parts act in concert with each other. For Turnbull, this revelation in physics confirmed what had been known going back to the ancients: that the moral world is a similarly coherent and law-abiding system. Thus, immediately after his encomium to Newton in the Preface, Turnbull states:

> Now, no sooner had I conceived this idea of moral researches [on a Newtonian model], than I began to look carefully into the better ancients (into Plato's works in particular) to know their opinion of human nature, and of the order of the world. And by this research I quickly found, that they had a very firm persuasion of an infinitely wise and good administration, actually prevailing at present throughout the whole of nature.

Although the ancients had not made great advances in natural philosophy, they *had* made great progress in comprehending the "moral constitution of man" along Newtonian lines.[20]

Later in the text, Turnbull points out that modern natural law theorists such as Hugo Grotius and Samuel Pufendorf had themselves drawn heavily from ancient writers; and throughout this and other works Turnbull joins the natural law tradition, stretching back to the ancients, to the discoveries and methods of the New Science.[21] The Roman writer Marcus Tullius Cicero (106–43 BCE) was by far Turnbull's favorite ancient writer on natural law, if we take the sheer number of times Turnbull cites him in *The Principles of Moral Philosophy* as evidence.[22] It cannot have been lost on Reid, under Turnbull's tutelage, that one could learn much from ancient writers about human nature and the constitution of the human mind and that there was no inherent or necessary tension between ancient and modern philosophy. If the world system was as orderly and law-abiding as Newton and others had shown, then genuine knowledge of our mind and constitution was not dependent on time and place; as Turnbull put it, "good reasons are good reasons" regardless of their provenance, as long as reasoning is based in fact, evidence, and careful analysis of experience rather than speculation.[23]

Thus when it comes to philosophical questions like the existence of liberty or necessity, everyone understands "what it is to be free, to have a thing in his power, at his command, or dependent upon him." It is only philosophers who theorize beyond experience that depart from such common understandings (i.e., who question whether we are truly free agents), "and therefore are not understood by others, and sadly perplex and involve themselves."[24] The assumption of an orderly and coherent world system thus inexorably leads to the idea that common understanding and universal consent are important forms of evidence about the nature of our minds; if philosophical conclusions conflict with them, we must be off-track.

Once Turnbull has laid the basic foundations of his enterprise, he proceeds to examine "the faculties and dispositions with which we are provided and furnished for making progress in knowledge"—that is, our mental furniture that makes all forms of knowledge possible. First, it is clear that the wisdom and goodness of nature provide us, from infancy, with the ability to insensibly and instantaneously form judgments "concerning such laws and connexions in the sensible world, as it is absolutely necessary to our well-being, that we should know [...] How soon do we learn to judge of magnitudes, distances and forms, and of the connexions between the ideas of sight and touch, as far, at least, as the common purposes and conveniences of life require." It is furthermore apparent that we have a wonderful facility to learn any language in our tender years; "but this most useful of all languages for us to know, the *language* of nature, as it may very properly be called, is what we learn soonest, and as it were necessarily and insensibly."[25] Human beings are thus naturally equipped to learn and understand both the language of other humans and the language of nature.

Later, toward the end of Volume II, titled *The Principles of Christian Philosophy*, Turnbull returns to this theme and adds further detail: "Indeed were we not capable, before we can reason, to form very quick and ready judgments of certain connexions in nature (concerning the magnitudes and distances of objects, for example) as we very early do, we could not possibly get thro' our infant state." The ability to make such judgments, which are in a sense "formed in us" as much as formed *by* us, "by the necessary operations of certain faculties belonging to us, previously to our use of reason [...] is a very manifest sign of the care of providence about us, whose reason must, in the nature of things, that is, according to our make, be gradually nursed and cultivated to any considerable degree of strength and vigor." Turnbull goes on to discuss the law-abiding powers of the association of ideas, memory, and habit, which build upon our prerational "quick and ready judgments" of natural connexions and magnitudes to form the bulk of our knowledge.[26]

Here Turnbull's thinking prefigures Reid's notion of principles of common sense that serve as a ground for higher knowledge; and it plainly outlines the way in which such a view ties common sense to science, as when Turnbull goes on to state: "For the connexions of nature lie open to every diligent judicious enquirer; every such a one is daily making, in proportion to his assiduity in observing nature, and trying experiments, very great discoveries with ease and pleasure." But given this fact, Turnbull asks, how is it that "natural science hath made such slow advances, and is yet so little studied and pursued[?]"

Turnbull's answer has a Baconian ring to it: "We shall [...] only observe upon that head, that in fact, philosophers were long misled from the plain and evident way of coming at the knowledge of nature [...] by a vain disposition, to make or contrive worlds themselves, and to spin a solution of all the phenomena of nature out of their own brain." It is thus vain, speculative philosophers, not common enquirers, who are out of touch with the progress of science. It is due to vanity, thoughtlessness, false notions of learning, and "sensuality" and the love of pleasure "that natural philosophy, the advantages of cultivating which, glare every thinking man in the face, is not yet pursued with earnest and assiduous application it ought to be."[27]

Returning to Turnbull's enumeration of our mental furniture in *The Principles of Moral Philosophy*, he lists, in addition to our natural ability to make quick and ready judgments about natural phenomena, a natural care and concern for our own preservation and well-being, as well as a "moral sense of beauty and deformity in affections, actions and characters, by means of which, an affection, action or character, no sooner presents itself to our mind, than it is necessarily approved or disapproved by us." Although it takes a long time for reason to mature, with our moral sense, "the Author of our nature has much better furnished us for a virtuous conduct than many philosophers seem to imagine, or, at least are willing to grant, by almost as quick and powerful instructions as we have for the preservation of our bodies." Here Turnbull cites Hutcheson, as he discusses the ways in which our moral sense directs our actions toward the "nobler pleasures" that intend the good of others while "undesignedly" promoting "our own greatest private good."[28]

We furthermore come equipped with the ability to form analogies and probabilities and to derive satisfaction and conviction from them, depending on the degree of likeness. One can call this effect of probability or likeness "a judgment or tendency to determine ourselves to act this or the other way, or not to act at all, according to the various force of presumption." And the only way to account for it is to suppose "an aptitude or disposition in our natures" to make such presumptions. The case is the same in perceptions of beauty, which in the end are affections of mind rather than results of rational analysis: "We must in all such cases at last come to an ultimate reason, which can be no other than the adjustment of the mind to certain objects."[29]

Turnbull goes on to consider two things "very remarkable in our nature": the association of ideas and the ability to form habits by repeated acts. "Unless the mind were so framed, that ideas frequently presented together to it, should afterwards naturally continue to recal one another, to blend or return together, habits could not be contracted." Thus "Those effects called the association of ideas and formation of habits, do therefore resolve themselves into the same general law or principle of our nature, which may be called the *law of custom*." This principle of our nature allows us to attain perfection in any science, art, or virtue. If we were not so constituted, "It would not be in our power to join and unite ideas at our pleasure, to recal past ones, or to lay up a stock of knowledge in our minds."[30]

However, given our natural propensity to join ideas together, it is the business of science to distinguish between true and false associations of ideas: we need to be

> incessantly upon our guard to prevent the blendings and cohesions of ideas, that the regular course of things in the world naturally tends [...] to form or engender in our minds. Everyone who is acquainted with philosophy, knows, that the great difficulty in attaining to the true knowledge of things, takes its rise from the difficulty of separating ideas into the parts that naturally belong to one another, from those which are added by association.[31]

As we will see, Reid developed a similar viewpoint on the role of scientific thinking, in his case by distinguishing between true and false attributions of causal connections in nature. For both thinkers, we are naturally framed to make connections between things that appear in our experience to be related to each another, and it is the job of science to distinguish true from false connections. That there is continuity between Turnbull and Reid on this general point is further indicated by statements Turnbull makes later on in the text, which sound very much like they could have been made by Reid. In discussing the "original sociality of our nature," for example, Turnbull suggests that to believe sociability is the effect of education, custom, and art, "without any intention or appointment of nature that it should be so, must terminate ultimately in saying, that effects may be produced without causes," which would "unanimously be owned to shock all common sense."[32]

Turnbull goes on to focus discussion on the moral sense, at times simply referring the reader to Hutcheson for further elaboration.[33] Thus although his primary interest is in moral philosophy and the moral sense, Turnbull's synthetic vision takes him into the wider precincts of a general philosophy of mind and how it connects with the world in all its aspects. For Turnbull, there is no strict dividing line between mental and moral philosophy, and between natural and moral science.[34] Analogies are everywhere. Just as we are framed to perceive and find pleasure in the modification of light and colors, so are we framed when it comes to our approval or disapproval of moral character. The question of a moral sense "is about a fact, a part of our constitution; about something felt and experienced within us, in consequence of our frame; and it cannot possibly be decided, but by consciousness, or by attending to our mind, in order to know how we are affected on certain occasions by certain objects."[35]

Given this fact of our nature, we can be sure that in questions of both visible appearances to the eye and moral appearances to the understanding, "where universally all languages make a distinction, there is really in a nature a difference." Here as elsewhere Turnbull develops the notion that a universal natural constitution necessarily implies that universal expressions of the human spirit—whether in language, oratory, or art—exist as factual evidence of universal laws and truths. Indeed, the common folk confirm the existence of a moral sense: "To be satisfied of the universality of this sense, let one but try the lowest of mankind in understanding, and fairly representing to him the virtues and vices, bring forth his natural, his first sentiments about them; for he shall find that even the most illiterate have a strong moral sense."[36]

However, to speak of a moral sense is not to suppose the existence of innate ideas, "it is only to assert an aptitude or determination in our nature to be affected in a certain manner so soon as they occur to the mind." Nevertheless, such a natural sense "is justly said to be engraven on our hearts, innate, original, universal." The "great business of reason" is to "cultivate, improve, and then preserve in due force this our rightly improved natural sense of right and wrong."[37] As Turnbull furthermore suggests in *The Principles of Christian Philosophy*, it is in vain to say that our moral sense "is totally acquired by reason, in proportion as it is improved [...] it cannot produce it when it is not originally implanted in some degree."[38]

As already alluded, Turnbull also argues that we have a natural sense of beauty; and such a sense of beauty is connected to utility: "When we come to examine those objects [of beauty] attentively," we find that there is "truth, proportion, regularity and unity of design to bring about [... an] advantageous end." Turnbull goes on to outline the notion that "the sound state" of something is "the beautiful one." In other words, our natural sentiments of beauty are perceptions of the goodness or health of that object in serving its end. Hence studying beautiful objects from all times and places helps us to identify goodness and truth as it exists in the world. Beautiful things are testimony to the harmonious order of both the natural and moral world, and there is much to be learned from contemplating them.[39]

This is the educational idea behind Turnbull's *Treatise on Ancient Painting*, also published in 1740 and containing many fine engravings of ancient works of art. In the "Epistle Upon Education" printed at the beginning of the *Treatise*, Turnbull repeats many of the views already discussed about the continuity between natural and moral science, with the specific aim of the book being to elucidate universal moral laws and principles from the consideration of ancient works of art. Turnbull observes that writings by traditional moralists like Grotius and Pufendorf are in essence dull and dry verbal affairs. Rather, "the properest way" of reasoning about morals using past authority is to examine proverbs and "images" (works of art) produced around the world, "whence it would appear how common, how universal good sense is, and always hath been."[40]

Thus if we are made by a God who superintends a rational, orderly world, there is no reason to suppose that human beings, as a creation of that same God, are not equipped to understand and make sense of that world, if only they understand the nature and right use of their faculties. Rational order and principles are endowed in all the creation, including in human nature; such endowments, which Reid would come to call principles of common sense, serve as the ground of all higher thinking. Turnbull, who as we have seen mentioned "common sense" occasionally if unsystematically, began to develop this idea but did not go into much philosophical depth. It was only with the emergence of Hume's renewed skeptical challenge that thinkers such as Reid would heed the call to delve deeper into the nature of the mind itself, shorn of overtly moral and theological considerations and focusing on the elemental processes by which we gain knowledge of the world. Turnbull applied scientific method in the service of moral philosophy, with important implications for education and the inculcation of public virtue.[41] Reid, in response to Hume, would go on to develop the implications of Turnbull's thought and the moral sense school into a true "science of the mind" that more deeply connected the mind, and indeed all minds, to the orderly and law-abiding world that was being described by modern natural philosophers.

* * *

There was, however, one writer preceding Reid, who, in responding to Hume, began to move the discussion further toward philosophical consideration of "common sense" in its own right. Henry Home, Lord Kames (1696–1782) was a quintessential Enlightenment figure. By profession a lawyer, he rose to become a leading judge and author, gentleman farmer, educator, literary critic, philosopher, and "improver" widely known to be a witty friend and companion. He was active in the philosophical and literary clubs that were sprouting up at the time, eventually becoming president of the Edinburgh Philosophical Society, and he did much to foster *Belles Lettres* in Scotland, both in his own writings (*Elements of Criticism*, 1762) and by encouraging the study of rhetoric and literature at the university in Edinburgh. Kames was friends with Benjamin Franklin and Thomas Reid (both of whom paid extended visits to his estate), David Hume, and Elizabeth Montagu, and he provided support and counsel to many others, including Adam Smith, Hugh Blair, James Beattie, James Macpherson, and James Boswell, who planned, but never completed, a biography of Kames.[42]

Kames wrote extended commentaries on legal decisions that helped to systematize Scottish jurisprudence (*Dictionary of Decisions*, 1741), but he also wrote light works on education (*Loose Hints upon Education*, 1781). There were highly regarded books on agriculture (*Progress of Flax-husbandry in Scotland*, 1766), historical anthropology (*Sketches of the History of Man*, 1774), and of most interest to us, moral philosophy and epistemology (*Essays on the Principles of Morality and Natural Religion*, 1751).

In all of his works Kames was concerned, in good lawyerly fashion, to elucidate the underlying principles and laws that order the particular facts of the case. Kames believed, as did Turnbull and a growing number of others, that human nature, as a subset of a providentially ordered universe, could be understood through empirical analysis, including analysis of one's own thoughts and feelings.[43] And one of the fruits of the scientific analysis of human nature was the finding that our senses—both external and internal— are authoritative and that furthermore "our reasonings on some of the most important subjects, rest ultimately upon sense and feeling."[44] This does not mean that our reasonings are arbitrary; quite the contrary, since sense and feeling are structured in an orderly fashion to achieve certain ends in the larger scheme of things. But what are the principles of sense and feeling? This was a question that, as we have seen, was gaining increasing currency at the time, ultimately leading to Reid's cataloguing of the "principles of common sense."

Kames, like Turnbull, asserted the existence of a number of feelings, sentiments, and principles of human nature—our "constitution"—that equip us for life in the world. For Kames they included, among others, a "sympathising principle"[45] that helps make us sociable; a "sense of duty" and "sense of justice" that do likewise; rules of conduct, approved by the moral sense—such as care for children, fulfillment of promises, gratitude, and benevolence; an idea or feeling of property that serves as a foundation of private property and is a component of our sense of justice; a sense or feeling that individuals are responsible for their actions; and a natural feeling or sense of moral liberty—a feeling that Kames deconstructs, in a Humean manner, as being illusory.[46] Kames thus did not suppose that just because we come into the world naturally constituted to feel and perceive things in a certain way, such feelings and perceptions are necessarily "true" in a strict sense. Since human beings are framed more as active than contemplative beings, we perceive the world in practical terms, rather than in theoretical ones, and hence "our perceptions some times, are less accommodated to the truth of things, than to the end for which our senses are designed."[47]

This is evident in regard to our feeling of liberty. On the one hand, careful attention to the human mind reveals a sentiment of causality that tells us effects have causes, and consequently that there are no uncaused events or actions. On the other hand, analysis also reveals that we have a sense or feeling of liberty; that we can initiate what are, in effect, actions that have no precipitating cause. Kames resolved this paradox by suggesting that "the truth of things is on the side of necessity; but that it was necessary for man to be formed, with such feelings and notions of contingency, as would fit him for the part he has to act."[48] In other words, our acts are not, strictly speaking, freely chosen, but we are so constituted to feel as though they are, in order to function as moral beings in society.

Our delusive feeling of liberty leads us to ascribe moral responsibility to each other, which helps to bring about moral behavior. Being the subject of moral disapproval, for example, might lead an individual to avoid such disapproval, even if that individual is not truly free to make moral choices. A man may steal out of need or greed, for example, but he may "choose" differently next time if the desire to avoid shame is added to his repertoire of competing motives.[49] And he will be more disposed to act responsibly if he believes he is free to choose his actions. If man were to understand himself to be simply part of a larger system of necessity, then inaction and passivity would be the result, putting an end to rational deliberation about the future, since there would be little reason to do so if there were no real choices to be made.[50] Therefore it is true enough for us, as rational actors who must get along in the world, that we are free agents, even if some of our other intuitive perceptions, when we reflect upon them, tell us that the system we are a part of must be determined.

Kames received much abuse for this argument from a couple of vociferous Scottish divines (until his major antagonist, the evangelical minister George Anderson, obligingly passed away in 1756), and he subsequently retreated from his original position, but the episode indicates that Kames's intent was not simply to defend religious orthodoxy against freethinkers like his friend David Hume. Rather, it shows that he was unafraid to develop an argument, based on facts and scientific analysis (and framed by a providentially designed universe), to its logical conclusion.

Kames's distinction between the cognitive requirements of everyday life and those of abstract philosophical reflection was an important element of common-sense philosophy, as was the relative weight given to everyday experience in deciding philosophical questions. In his view, Kames was simply establishing what appeared to be an indubitable scientific fact of human nature—we are constituted to "make sense" of the world in ways that ground our experience and reasoning about that world—which has important consequences for how we decide philosophical questions, including the problems posed by modern idealism and skepticism.

Kames took up these issues in Part II of the *Essays*, which contain the earliest expression of a consciously developed philosophy of common sense published in Britain. Kames lays out in rough detail a number (but by no means all) of the basic philosophical positions that Reid would go on to develop in much more detail and with considerably more subtlety. Although it would be wrong to overemphasize the debt which Reid, who in later years visited Kames during summer holidays and corresponded with him at other times, owes to Kames, it would also be a mistake not to recognize the degree to which Kames's thought prefigured and laid further groundwork for Reid's philosophical achievement. Published 13 years before Reid's *An Inquiry into the Human Mind, On the Principles of Common Sense* in 1764, the *Essays* offer a coherent (albeit sketchy) common-sense philosophical response to the skeptical, idealistic, and reductionistic trends of modern philosophy. Kames thus further extended the process, begun by Turnbull, of expanding the moral sense position into metaphysics and epistemology.[51]

Kames begins Part II by discussing the nature of belief. Rejecting Hume's contention that belief is simply a lively conception of an idea, Kames argues that belief is "founded upon the authority of our senses," either mediately (via the testimony of others) or immediately (via our own sense experience). We are constituted by nature to give sense information an "irresistible authority," just as we are constituted normally to tell the truth and believe that others are doing the same.[52] Reid would go on to call these latter endowments the "principle of veracity" and the "principle of credulity" and argue that they are essential building blocks of social intercourse.[53]

When it comes to the idea of a coherent self or personal identity, the reasonableness of which Hume had questioned, Kames suggests that human beings have "an original feeling or consciousness" of themselves that is self-evident to common understanding and in no need of philosophical explication. If such "natural feelings, whether from internal or external senses, are not admitted as evidence of truth, I cannot see that we can be certain of any fact whatever."[54] If we cannot be assured of a coherent self that experiences the world, then it is not possible to be assured of the coherence of any particular facts that the self observes in the world. Here philosophical speculation runs up against psychological analysis of the sense-making it takes to "be in the world."

Even though our senses can be deceived, "there is nothing to which all mankind are more necessarily determined, than to put confidence in their senses"; it is simply not in our power to doubt them. Things are not necessarily as they appear to us (as in the case of secondary qualities such as colors or sounds), but a little reflection can easily disabuse us of false appearances; and when we are deceived, there is usually some useful purpose to it. There is a *reason*, confirmed by constant experience, that we are disposed to put confidence in our senses. It is thus strange

> that it should come into the thought of any man to call [the perceptions of external objects by our senses] in question. But the influence of novelty is so great; and when a bold genius, in spite of common sense, and common feelings, will strike out new paths to himself, 'tis not easy to foresee, how far his airy metaphysical notions may carry him.

Here Kames is referring to Bishop Berkeley's denial of the independent reality of external objects (discussed in Chapter 2), a denial that Kames believes leads to total skepticism: "If we can be prevailed upon, to doubt of the reality of external objects, the next step will be, to doubt of [...] the reality of our ideas and perceptions. For we have not a stronger consciousness, nor a clearer conviction of the one, than of the other."[55]

The basic conflict is between discursive reason and the deliverances of our senses and feelings: to which are we going to give priority? The common-sense position is that it is wrongheaded to discount the latter in favor of the former, when both, as Reid would later say, "came out of the same shop."[56] Both are natural endowments with particular roles to play in our sense-making of the world. It is *reasonable*, most of the time, to trust our senses, even if abstract *reasoning* calls what they tell us into question. It simply isn't possible rationally to prove or demonstrate every proposition that we feel or perceive to be true—this is one of the constant drumbeats of Kames's *Essays* and of common-sense philosophy in general. To call common-sense philosophy "a kind of dogmatic anti-intellectualism"[57]—as does one of Kames's modern biographers—simply indicates the depth of the challenge it poses to the rationalistic tendencies of much of modern philosophy, rather than any inherent weakness in the argument.

In Kames's view, "common sense and experience" are the best assurances we have of the existence and reality of external objects. And one reason it is necessary to unravel the arguments of skeptics and idealists that undermine such data is to show "that our senses, external and internal, are the true sources, from whence the knowledge of the Deity is derived to us."[58] The apologetic dimension in Kames's writings—as in Turnbull's—is clear; it would recede into the background in Reid. The existence of such an intent, however, says nothing about the quality of the argument or the many other uses to which it was put. The way forward, at any rate, was "a distinct analysis of the operations of those senses, by which we perceive external objects."[59] Such an analysis of the five external senses would become the major task of Reid's *An Inquiry into the Human Mind, on the Principles of Common Sense.*

Kames goes on to make his own cursory analysis of the senses, finding that "independent and permanent existence" is bound up with our perception of external objects.[60] When I lay my hand upon a table, I receive an impression of a hard, smooth body that creates not the least suspicion of fallacy; and when we perceive objects by sight, nature conceals the physical impression of light hitting the retina, "in order to remove all ambiguity, and to give us a distinct feeling of the object itself"; if we felt the impression on the retina we would want to place the visible object there.[61]

As for our idea of causal power, "Sense and feeling afford me a conviction, that nothing begins to exist without a cause, tho' reason cannot afford me a demonstration of it." The idea of power is not deducible from experience; all we can gather from experience, as Hume had shown, was an idea of constant conjunction, "which comes far short of our idea of cause and effect." Rather, causality is a spontaneous perception we have when we observe physical events like one billiard ball crashing into another:

We are obviously so constituted, as not only to perceive the one body acting, and exerting its power; but also to perceive, that the change in the other body is produced by means of that action or exertion of power. This change we perceive to be an *effect*; and we perceive a necessary connection betwixt the action and the effect, so as that the one must unavoidably follow the other.

Such feelings and perceptions cannot be explained, other than by suggesting the terms that denote them and asking others to attend to their own perceptions in similar circumstances. Without such spontaneous, inexplicable perceptions we could never have the idea of body or its powers—neither reason nor experience can give us such perceptions.[62]

It is true that we err by the misattribution of causes to effects; yet the faculty that assigns causes to effects is exerted with sufficient certainty "to guide us through life, without many capital errors."[63] As Kames states in *The Art of Thinking*, when we see stacks of coal in front of peasants' houses, we can be sure coal is abundant in the area, even if in other instances causal inferences can be mistaken.[64] Whereas, according to Kames, Hume had claimed—in opposition to the common sense of humankind—that the power or energy of causes "belongs entirely to the soul," Kames argues that causal powers are spontaneously perceived as qualities located in the acting bodies and are by no means reducible to mere operations of mind. Just because we are not able to give a clear explanation of how this happens, it is still an empirical fact that cannot be denied, especially in light of "the wonderful harmony that exists [between the feeling of power] and the reality of causes and their effects."[65] The correspondence between our sense of power or agency and the way the world actually works, in our experience, is strong evidence that our perceptions are accurate. Causality is not just in our heads, but clearly in the world as well.

In addition to the perception of causal power, we are constituted to perceive nature as stable, coherent, and uniform over time—in other words, that the future will resemble the past. "This perception or feeling must belong to an internal sense, because it evidently has no relation to any of our external senses."[66] Reid would later call this the "inductive principle" and argue that without it all science as well as the common activities of everyday life would be impossible.[67] As Kames saw it, this principle is so strong that it makes us look for constancy and uniformity even when our experience suggests otherwise— the human tendency, highlighted by Bacon, of seeing patterns and connections even where there are none to be found.

In sum, the Author of our nature has done two things, according to Kames: "He has established a constancy and uniformity in the operations of nature. And he has impressed upon our minds, a conviction or belief of this constancy and uniformity, and that things will be as they have been."[68] Here Kames steps back to contemplate the theological frame that encloses his argument, and the *Essay* concludes with a discussion of how the existence of the Deity is attested by our perception of causality. Such wondrous effects as we see in the world suggest a "cause, unbounded in power, intelligence, and goodness."

Kames's friend David Hume saw the matter differently, of course. Hume had been led to the conclusion that reason was impotent and unable, on its own, to demonstrate order in nature, and hence the existence of a deity. Kames was well aware of the force of Hume's arguments, and he shared Hume's strictures against the productive power of discursive reason.[69] Yet in Kames's view, Hume himself put too much stock in reason "when he undertakes to argue mankind out of their senses and feelings."[70] That is to say that Hume's reasonings on such things as our perception of causality went against what everyone "knows" via normal sense experience—for example, that every effect has a cause—and there is something deeply problematic with this: namely, that such basic perceptions and feelings provide the basis for rational thinking in the first place. This was a theme that Reid would take up and develop at length, but it is already present in Kames's *Essay*.

Kames's social outlook corresponded to his philosophical views. He was not only keen to stand up for "the common sense of mankind"[71] against the paradoxes of philosophers, but he also believed in the equality of human beings "not [...] in ability, character and outward circumstances of fortune, but in a moral sense and before the law, regardless of station, occupation, sex, and whatever environmental circumstances and education may had made of them."[72] In *The Art of Thinking* Kames repeatedly gives voice to this perspective with maxims like "The Prince, so magnificent in the splendour of a court, appears behind the curtain but a common man" and "Wisdom is better than riches; nevertheless the poor man's wisdom is despised, and his words are not heard."[73]

However, like most Enlightenment figures of his time, there were limits to Kames's democratic impulses—he was not a "leveller" and he occasionally spoke of the "vulgar" or common people in less than flattering terms.[74] A defense of common sense did not entail an uncritical acceptance of common opinion or the thought processes of the uneducated person. In *The Art of Thinking* Kames suggests that "Not one of a thousand thinks for himself; and the few who are emancipated, dare not act up to their freedom, for fear of being thought whimsical."[75] But if "a plain man, sincere and credulous, will build upon very weak testimony," the "diffident and suspicious will scarce be satisfied with the strongest. It is the province of reason and experience to correct these extremes."[76]

Above all, according to William Lehmann, "[Kames] was insistent that before the law, as before the moral judgement seat, there be no respecting of persons." Whether a case involved the Douglas clan or a common criminal, Kames "asked not for pedigree or tax-roll or place of abode, but merely for the facts in the case, and logical conclusions to be drawn from the facts, and the specific provision of the law that needed to be applied—nothing more."[77] Thus if common-sense philosophy was religiously conservative, it could also be socially progressive. By according philosophical validity to the ordinary perceptions of average human beings, it pointed toward the democratic future and away from the aristocratic past. If his friend Benjamin Franklin was the aristocratic democrat, "Kames was the democratic aristocrat."[78]

* * *

Turnbull and Kames, each in their own way, helped lay the intellectual foundations for a philosophy of common sense shorn of overtly moral and theological concerns. Turnbull's philosophy demonstrates quite clearly the scientistic pedigree of common-sense philosophy, while Kames delved more deeply and directly into "principles of common sense" as part of the emerging response to Hume and the project of applying scientific method to the study of the mind—all of which occurred in a theological context, to be sure. But together with the advances and discoveries of modern science, the providentially ordered world of Hutcheson, Turnbull, and Kames provided an intellectual framework for constructing a holistic philosophy of mind that reconnected "scientific" thinking to the wider world, including an emerging commons that played an increasingly important role in modern life. It is thus to Reid's *rapprochement* between common sense, science, and the public sphere that we now turn.

Chapter 8

COMMON SENSE, SCIENCE, AND THE PUBLIC SPHERE: THE PHILOSOPHY OF THOMAS REID

The thought of Thomas Reid (1710–1796) presents a fitting climax and culmination of the story we have been following about the relationship between common sense and science dating back to the ancient Greeks. As we have seen, by the turn of the eighteenth century the New Science had challenged the intellectual primacy of common-sense experience in favor of recondite, expert, and even counterintuitive knowledge increasingly mediated by specialized instruments. Meanwhile modern philosophical thinking that emerged in tandem with the New Science, including skeptical and materialist currents of thought, had problematized the perceptions of everyday sense experience and accepted understandings of the self, morality, religion, and society. Everything was "up for grabs," intellectually, and thinkers from across the intellectual spectrum and throughout Europe felt compelled to revisit the grounds of our most basic beliefs and forms of knowledge.

In the case of Reid's contemporary David Hume (1711–1776), reexamination of the fundamental bases of human knowledge led to skeptical scrutiny of, among other things, our perception of causal relations in nature, a fundamental precondition of scientific endeavor. In this chapter I argue that in responding to this "problem of induction" as advanced by Hume, Reid completed the long-term philosophical process, outlined in previous chapters, of reconnecting everyday understanding and experience with the findings and methods of modern natural science. An educator and mathematician self-consciously working within the framework of the New Science, Reid articulated a philosophical foundation for natural knowledge anchored in the human constitution and in processes of adjudication in an emerging modern public sphere of enlightened discourse. Reid thereby completed the ongoing intellectual transformation of one of the bases of Aristotelian science—common experience—into a philosophically and socially justified notion of "common sense." This fact, along with the fact that his perspective—and that of the common-sense/moral sense philosophical tradition generally—was supportive of orthodox Christian theology and moral precepts, helps to explain why Reid's thought was so influential in nineteenth-century Europe and America (discussed at length in the Epilogue).[1] Yet whatever its theological and moral dimensions, Reid's thinking was centrally concerned with epistemology and the philosophy of science[2] and was bound up with wider sociological, philosophical, and scientific changes taking place in the early-modern period.

* * *

Thomas Reid was born in 1710 in Strachan, Kincardineshire. His father was a minister in the Church of Scotland, and his mother, Margaret Gregory, was from a prominent family whose members included James Gregory, inventor of the reflecting telescope, and David Gregory, Savilian Professor of Astronomy at Oxford and a friend of Isaac Newton. Two of her brothers were professors of mathematics and among the first to teach Newton's ideas in Scotland. Reid entered Marischal College in Aberdeen at the age of 12, where he became a pupil of George Turnbull, who as we have seen was a strong proponent of Newtonian method. Reid thus came by his scientific/philosophical proclivities both genetically and by way of education and culture.

Reid graduated MA from Marischal College in 1726 and went on to study divinity at Marischal until 1731, when he was licensed to preach by the presbytery of Kincardine O'Neil. From 1733 to 1736 he was librarian at Marischal, which gave him time for study, and in 1737 he was ordained as minister of the parish of New Machar, about 10 miles to the northwest of Aberdeen, where he remained until 1751, when he was elected Regent at King's College, Aberdeen. Reid taught at King's until 1764, when, upon the publication of *An Inquiry into the Human Mind, on the Principles of Common Sense*, he was elected Professor of Moral Philosophy at Glasgow University, succeeding Adam Smith. In 1780 he began to step back from teaching, and he spent the following years publishing a systematic account of his thinking in two volumes, *Essays on the Intellectual Powers of Man* (1785) and *Essays on the Active Powers of Man* (1788). Reid was active in a number of extracurricular societies, including the Aberdeen Philosophical Society, the Glasgow Literary Society, the General Assembly of the Church of Scotland, and the Highlands and Islands Commission, among others. He and his wife Elizabeth had nine children, of which only one, Martha, lived past the age of 30. Reid himself lived a relatively long life, by the standards of his day—he died at the age of 86 in 1796, his life having spanned a good part of the eighteenth century.[3]

A short essay entitled "Of Common Sense," published in 1785 in *Essays on the Intellectual Powers of Man*, provides a useful point of entrée into Reid's general philosophical perspective, as well as his understanding of his place within the history of ideas. Reid begins the essay by arguing that modern philosophers such as Locke have been mistaken in defining "sense" as having nothing to do with judgment. This is due to the Lockean "theory of ideas," which holds that sense is "the power by which we receive certain ideas or impressions from objects; and judgment as the power by which we compare those ideas, and perceive their necessary agreements and disagreements." Yet in common language "sense always implies judgment. A man of sense is a man of judgment [...] Nonsense is what is evidently contrary to right judgment." Thus, "Common sense is that degree of judgment which is common to men with whom we can converse and transact business." Whereas philosophers understand seeing and hearing to be senses in that they provide ideas to the mind, "by the vulgar they are called senses, because we judge by them. We judge colours by the eye; of sounds by the ear; of beauty or deformity by taste; of right and wrong in conduct, by our moral sense, or conscience." Philosophers themselves, in unguarded moments, "fall unawares into the popular opinion" that senses are judging faculties; and this popular understanding of sense is not confined to the English language but is evident in Greek, Latin, and all European languages.[4]

Reid thus grounds his analysis of the meaning of "sense" inductively and pragmatically, in common transactions, language, and opinion, finding the notion to be laden with judgment and not simply a producer of data for the mind to analyze. He goes on to argue that he can find no reason why philosophical uses of the term should depart from the commonly accepted meaning. Quoting Alexander Pope's phrase that "good sense" is a "gift of heaven," Reid goes on to say that such an "inward light or sense" is given to different persons in different degrees, and a certain degree of it is necessary to us as subjects of laws and government and in the management of our own affairs. The laws of all civilized nations distinguish between those that have this gift of heaven from those that do not, and it is easily discernible by judges and juries.[5]

Reid then moves on to suggest that "The same degree of understanding which makes a man capable of acting with common prudence in the conduct of life, makes him capable of discovering what is true and false in matters that are self evident, and which he distinctly apprehends." Further, "All knowledge, and all science, must be built upon principles that are self-evident; and of such principles, every man who has common sense is a competent judge, when he observes them distinctly." Rational discussion is based on agreed first principles, and "when one denies what to the other appears too evident to need, or to admit of proof, reasoning seems to be at an end; an appeal is made to common sense, and each party is left to enjoy his own opinion."[6] These are all very common themes in Reid's writings.

Concluding his opening discussion, Reid reiterates the notion that sense, "in its most common, and therefore its most proper meaning, signifies *judgment*, though philosophers often use it in another meaning. From this it is natural to think, that common sense should mean common judgment; and so it really does."[7] Reid avers however that even if we can all agree on the meaning of the term, setting the precise limits dividing common judgment from what lies beyond it may be difficult to determine; and indeed he spends a great deal of effort in his works delineating the true principles of common sense, as distinct from higher forms of knowledge that are derived from them (discussed further below).

Thus "common sense" is a term as well understood as "county of York"; most people know what is meant without necessarily being able to define its precise limits. The term is to be found in innumerable places in "good writers" and is heard often in conversation with the same general meaning.[8] Reid goes on to provide a brief survey of uses of the term by philosophers and other writers, including Dr. Johnson, Buffier, Berkeley, Fénelon, Hume, Cicero, and the Third Earl of Shaftesbury. He notes that the term is in many cases not systematically used by philosophers, but there are exceptions including Buffier and Berkeley. Buffier for his part "treated largely of common sense, as a principle of knowledge, above fifty years ago."[9] And despite Reid's critique of Berkeley's general philosophical perspective (see Chapter 2), here Reid credits Berkeley as one who "laid as much stress upon common sense, in opposition to the doctrines of philosophers, as any philosopher who has come after him."[10]

It is the thought of Shaftesbury, however, that merits the bulk of Reid's attention in the short essay and is clearly a touchstone for Reid's general perspective on common sense. Reid summarizes Shaftesbury's *Sensus Communis: An Essay on the Freedom of Wit and Humour* at length, before concluding that Shaftesbury had a double intention in the *Essay*: one to justify the use of wit, humor, and ridicule among friends when discussing "the gravest subjects"; the other to show "that common sense is not so vague and uncertain a thing" as skeptics think. Shaftesbury further shows in a "facetious way," throughout the *Essay*, "that the fundamental principles of morals, of politics, of criticism, and of every branch of knowledge, are the dictates of common sense." And he "sums up the whole in these words: 'That some moral and philosophical truths are so evident in themselves, that it would be easier to imagine half mankind run mad [...] than to admit any thing as truth, which should be advanced against such natural knowledge, fundamental reason, and common sense.'"[11]

Reid goes on to quote similar-sounding passages from Françoise Fénelon, Archbishop of Cambrai (discussed in Chapter 5), in a similarly approving fashion, suggesting that Fénelon's commonsensical gloss on the Cartesian criterion of truth "is the most intelligible and most favorable I have met with."[12] Reid then discusses the use of the term that he believes supports his position that is found in Cicero, Hume, and Joseph Priestley (who was a vehement critic of Reid), before concluding that "The authority of this tribunal is too sacred and venerable, and has prescription too long in its favor to be now wisely called into question."[13] Reid thus locates himself in the longer tradition of thinking about self-evident first principles that has been the subject of this book and cites it as evidence for the existence of universally shared first principles of knowledge.

Elsewhere in the *Essays on the Intellectual Powers of Man* he provides a more panoptic view of the progress of philosophy:

> The remains of ancient philosophy [of mind], are venerable ruins, carrying the marks of genius and industry [...] In later ages, Des Cartes was the first that pointed out the road we ought to take in those dark regions. Malebranche, Arnauld, Locke, Berkeley, Buffier, Hutcheson, Butler, Hume, Price, Lord Kames, have laboured to make discoveries; nor have they laboured in vain. For, however different and contrary their conclusions are, however sceptical some of them, they have all given new light, and cleared the way to those who shall come after them.[14]

Thus despite his often trenchant criticism of the "theory of ideas" that had arisen in the wake of Cartesian skepticism,[15] Reid felt that philosophers like Descartes, Locke, and Hume occupied an important place in the history of ideas, alongside more amenable thinkers like Buffier, Hutcheson, Butler, and Kames.

Reid proceeds to draw together the various threads of the short essay with some concluding comments. He states that it is an absurdity to conceive of any opposition between reason and common sense; indeed, "[common sense] is the first born of reason, and as they are commonly joined together in speech and in writing, they are inseparable in their nature." Thus we ascribe to reason two "offices" or "degrees": the first is to judge of things that are self-evident, and the second is to "draw conclusions that are not self-evident from those that are." The first is the sole province of common sense, "and therefore it coincides with reason in its whole extent, and is only another name for one branch or one degree of reason." However, if common sense is merely a degree of reason, why does it need a particular name? Reid's reply is to ask why one would abolish a term that is found in the language of all civilized nations and thus "has acquired a right by prescription?" There must be some specific use for the term; and there is indeed an obvious reason for it and that is "in the greatest part of mankind no other degree of reason is to be found" and is what entitles them to be called reasonable creatures. "It is this degree of reason, and this only, that makes a man capable of managing his own affairs, and answerable for his conduct toward others."[16]

This first degree of reason "is purely a gift of Heaven" and cannot be taught. "The second is learned by practice and rules, when the first is not wanting. A man who has common sense may be taught to reason. But if he has not that gift, no teaching will make him able either to judge of first principles or to reason from them." Furthermore, "the province of common sense is more extensive in refutation than in confirmation." What this means is that common sense only has jurisdiction to judge of reasoned conclusions that contradict it. It cannot confirm just reasonings from true first principles since the conclusions go beyond common sense, but it can recognize and judge a conclusion that contradicts "the decisions of common sense" by setting out from false principles or via an error of reasoning, even if it is not able to show the error of reasoning that led to the false conclusion. "Thus, if a mathematician, by a process of intricate demonstration, in which some false step was made," concludes that two quantities equal to a third quantity are not equal to each other, "a man of common sense, without pretending to be a judge of this demonstration, is well entitled to reject the conclusion, and to pronounce it absurd."[17] Thus ends Reid's essay "Of Common Sense."

I have dwelt at length on this short essay because it encapsulates major features of Reid's thinking about first principles of common sense and their relation to the wider world, along with his understanding of his own place in the history of ideas. As we move deeper into his thought, it will be helpful to remember that for Reid common sense is a species of judgment possessed by all healthy human beings in all times and places and, as such, is expressed in all languages. It is the basis for "self-evident" knowledge, that is, what is spontaneously known by all healthy human beings in their normal transactions with other humans and the world at large, and is a basic form of reason. It gives us information about fundamental features of that world—including other humans—and in so doing provides the basis for all higher forms of rational thinking (including natural science), as well as for navigating the practical affairs of life. As such, Reid's discussion of the relationship between common sense and higher forms of reasoning prefigures Daniel Kahneman's distinction between our mental "System 1" and "System 2," discussed in the Introduction.

Reid's thought was thus deeply grounded in the world of material things, other human beings, language, and society. And as we have seen, it picks up themes and ideas that have been present throughout history and that gained particular salience in the modern period as scientific thinking and the philosophy that grew in its wake began to drive a wedge between common sense and science.

* * *

Reid was not particular about the terms he used when talking about principles of common sense. In the *Essays on the Intellectual Powers of Man*, for example, he notes that when "propositions which are no sooner understood than they are believed" are "used in matters of science, [they] have commonly been called *axioms*; and on whatever occasion they are used, are called 'first principles,' 'principles of common sense,' 'common notions,' 'self-evident truths.' Cicero calls them 'naturae judicia' [...] Lord Shaftesbury expresses them by the words, 'natural knowledge,' 'fundamental reason,' and 'common sense.'"[18]

Reid goes on to distinguish between two kinds of self-evident truths: contingent and necessary truths. The former correspond to matters of fact, and as a result of reasoning yield probable conclusions, while the latter correspond to abstract truths and yield necessary conclusions. First principles of contingent truths include (but are not limited to) the following: (1) "The existence of everything of which I am conscious"; (2) that the thoughts I am having are my thoughts; (3) the things that I distinctly remember did really happen; (4) immediate and unreasoned knowledge of my personal identity and continued existence; (5) "that those things do really exist which we distinctly perceive by our senses, and are what we perceive them to be"; (6) "that we have some degree of power over our actions, and the determinations of our will"; (7) that our natural faculties "by which we distinguish truth from error are not fallacious"; (8) "there is life and intelligence in our fellow man with whom we converse"; (9) that sounds of voice, gestures, and countenance "indicate certain thoughts and dispositions of mind"; (10) that there is a reason to believe human testimony in matters of fact and even in matters of opinion; (11) that there are many events that are dependent upon human will; and (12) "in the phenomena of nature, what is to be, will probably be like what has been in similar circumstances." This self-evident principle that the future will resemble the past is a principle of our constitution that is confirmed by reasoning, which also makes us cautious in its application, helping us distinguish "accidents" from laws of nature. "In order to this, a number of experiments, varied in their circumstances, is often necessary." While all of the above principles are relevant to natural science, this last one is the most obviously important.[19]

When it comes to first principles of necessary truths, they include: (1) some grammatical principles like every adjective belongs to a substantive; (2) certain logical axioms, for example, no proposition can be both true and false, or that circular reasoning proves nothing; (3) mathematical axioms (e.g., Euclidean geometry); (4) axioms in matters of taste, such as "fundamental rules of poetry, and music, and painting"; (5) certain judgments of the moral sense or conscience, like "an unjust action has more demerit than an ungenerous one"; and (6) "metaphysical" first principles such as the fact that qualities perceived by our senses must have a subject or that whatever begins must have a cause that produced it.[20]

About 15 years earlier, in an unpublished essay most likely written for presentation to the Glasgow Literary Society, Reid had discussed the activity of our faculty of common sense thus: "By that faculty which we call common sense we compare objects that are presented to us and discern various affections and relations belonging to them, things concerning them, such as identity or diversity, number, similitude, contrariety, proportion, sum, difference, quantity, quality, time, place, genus and species, subject and accident, whole and parts and innumerable other relations." We furthermore have an immediate conviction of our own continued identity, and we have intuitive notions of identity and diversity in general, which are not subject to further analysis. Extension and figure are primary notions that we have via sense. Indeed, there are "large classes of notions" in every art and science that are clearly apprehended by common sense.[21]

Thus for Reid there are many first principles arising out of our faculty of common sense—including the principle of causality—that are simply taken for granted in our experience and thus cannot be demonstrably "proven," spanning all aspects of our world and helping us to make sense of that world. In the Epilogue we will pause to consider what modern scientific research tells us about what Reid got right, when it came to identifying first principles of common sense. Suffice it to say here that he does seem to have been correct on a number of points, including his assertions about the roots of our perception of cause and effect, a topic to which we now turn.[22]

* * *

Reid's seminal text, *An Inquiry into the Human Mind, on the Principles of Common Sense*, was published 20 years before his *Essays*, in 1764. This penetrating and influential work continues to garner the attention of philosophers and students of the human mind.[23] Since it is beyond the scope of this chapter to present a comprehensive analysis of the *Inquiry* and his other writings, I will focus on Reid's dispute with David Hume over the nature of our perception of causality as the most salient point of departure for exploring Reid's role in our story.

In his *Treatise of Human Nature* (1739–40) David Hume had argued that our mind operates by associating "ideas" derived from impressions of sensation or reflection in three basic ways: by the resemblance between ideas, by the contiguity of ideas, and by cause and effect.[24] Cause and effect is the most extensive mode or relation of the association of ideas, according to Hume (12). Hume argues that our notion of cause and effect cannot be gotten from abstract reasoning or reflection, rather it comes from our cumulative experience of observing the constant conjunction of objects in space and time (77). Our notion of causality is not intuitively or logically necessary, so it must be based on experience (82) and on the work of the imagination in joining events together into a seemingly necessary connection (93). Thus, belief in matters of fact and cause and effect arises from custom, habit, and experience, rather than any rational deduction or intuition from the observation of phenomena. We perceive distinct existences, "but no connexions among distinct existences are ever discoverable by human understanding. We only *feel* a connexion or determination of the thought, to pass from one object to another" (635). What has come to be called "the problem of induction," then, is the problem of how we get from particular perceptions to universal knowledge claims about the world, including those of causal relations, if we only have particular perceptions.[25]

Hume's skepticism about our rational faculties entailed a world in which human actors, including philosophers but especially the common people, were deceived as to the sources of their knowledge, belief, and action. Thus, for example, the "vulgar" tend to "confound perceptions [… with the] objects [which they represent]" (193), in common life we are under an "illusion" about cause and effect (267), and the "vulgar" particularly are misguided in applying their rules (150); "superstition arises naturally and easily from the popular opinions of mankind" (271), and so on.[26] As we shall see, although Reid agreed with Hume that superstition was the result of improper popular *attributions* of cause and effect, he maintained that the common people were in an important sense *not* deluded in their fundamental, universal understandings, like that of an intrinsic relation between an observed effect and a prior causative event. Hume was thus more pessimistic about the possibility for rational action in the public sphere than was Reid, and he exhibited a disdain for what he perceived to be the ignorance and lack of civility of common people not found in Reid.[27]

Hume's strictures on our perception of causality might have seemed less shocking to Reid had it not been for the fact that they represented an attack not only on the rational basis of everyday perceptions but on scientific knowledge as it had been understood since Aristotle (see Chapter 1). As R. S. Woolhouse notes, *scientia* in its Aristotelian sense meant "knowledge which, by virtue of the way it is structured, gives an understanding of why certain facts necessarily are so. This is done by explaining them in terms of their 'causes.'" In fact, it is somewhat redundant to speak of "causal explanation" since for Aristotle "an 'explanation' shows that something is so because of something else, and a 'cause' was taken to be whatever followed any 'be*cause*.'"[28] In the *Metaphysics* Aristotle equates wisdom with knowledge of these "becauses,"[29] and his definition of scientific knowledge in the *Nicomachean Ethics* states further that having such knowledge involves the capacity to demonstrate universal and necessary connections between things, proceeding via induction from known starting points to the generation of universal truths with the aid of syllogism.[30]

Granting a certain degree of variability in just what the notion of "cause" entailed in the ancient world,[31] these assumptions survived intact through the Middle Ages, and causal explanation remained, with some modification, the aim of Baconian natural philosophy.[32] Even Hume assumed that *scientia* entailed systematic knowledge of necessary connections.[33] Nevertheless, as we have seen in Chapters 2, 3, and 4, an epistemological erosion had long been underway in the early-modern period that culminated in Hume's rejection of the rational basis of our perception of causal relations in nature, now understood primarily as a function of *efficient* causes. Knowledge based solely on sense experience had long been recognized as inherently fallible,[34] and the rise of nominalism and skepticism in the early-modern period had led, broadly speaking, to a search for the foundations of certainty within the self and the workings of the mind.[35] But knowledge of the world predicated on reason and/or intuition alone also became suspect, importantly in Locke's thinking, leading to the acceptance of contingency and probability as the best we can do when it comes to natural knowledge. Since knowledge was seen to be mediated by simple "ideas" arising from experience, direct access to fundamental realities, including causes, was denied by Lockean epistemology.[36] In Locke's words, "Though causes work steadily, and effects flow constantly from them, yet their connexions and dependencies being not discoverable in our ideas, we can have but an experimental knowledge of them." Our ignorance of causes means that we are incapable of "a perfect science of natural bodies."[37]

Given this background, Hume is still typically seen in the way that Reid saw him: as building on the skeptical and nominalist foundations of his early-modern predecessors, pursuing skeptical arguments to their logical conclusion,[38] and arguing (in Stephen Buckle's words) "that *scientia* is impossible for creatures like us, and therefore that we are not to be distinguished from animals by any capacity for rational insight into nature."[39] To be sure, Hume took the existence of causal relations in nature for granted; the main thrust of his discussion of causality was to question its *intelligibility*.[40] But although even in Hume's thought realist presuppositions remain, the scope of human knowledge of nature has been severely limited. The "realist commitments" of the British tradition of early-modern philosophy appear to be threatened,[41] and the intelligibility of causal relations in nature has been seriously called into doubt.

Thus despite the fact that contemporary scholarship has identified a constructive side to Hume's thought,[42] Hume was almost universally perceived to be an epistemological and hence moral skeptic by his contemporaries, Reid included, and there were pretty good reasons for feeling that way.[43] But Reid responded not simply as a moralist, shocked at the debilitating effects that Humean ideas might have on religious beliefs and moral behavior. What Reid felt to be truly shocking was Hume's assault on common features of the understanding—"principles of common sense"—and the threat it posed to a scientific understanding of the mind and to the foundations of our knowledge of the natural, as well as moral, world.[44]

In the dedication to Reid's *Inquiry*, Reid states that Humean skepticism "is not more destructive of the faith of a Christian, than of the science of a philosopher, and of the prudence of a man of common understanding."[45] That is to say that Hume's brand of skepticism impinges not only on religious belief but also on rational, scientific thinking and on everyday understandings expressed in the common affairs of life. Reid's reply to Hume draws upon these three forms of understanding in the *Inquiry*, but his theism takes a back seat to inductive science and common understanding in arguing his case. God is the ultimate source of the order that we find in ourselves and in nature, but Reid is seeking to be a "scientific" philosopher and he engages the issues in an inductive, empirical, and naturalistic manner throughout his works.[46]

Reid's divergence from Hume is clear from the outset, when he indicates that his researches will reveal what every man feels and perceives (5), that observation and experiment, being the only way to gain knowledge of nature, are familiar to every human being in the common affairs of life and that in fact Newton's *regulae philosophandi*[47] "are maxims of common sense, and are practiced every day in common life" (12). In Reid's view, Hume had promised a complete system of the sciences built upon the foundation of human nature, yet "the intention of [his] whole work is to show, that there is neither human nature nor science in the world" (20). Reid's response was to write an inductive treatise, dividing the *Inquiry* into five parts, with chapters on each of the five senses. What he concludes from a thorough empirical investigation of the senses is that in their basic functions they give the world to us, and through them our mind grasps the world, in irreducible ways, certainly not reducible to Humean terminology. Once we are aware of the complexities of sensation and perception, we will find that there can be no other reasonable explanation for our basic processes of cognition than that we are set up or constituted in that way—nowadays we might say "hardwired"[48]—and that we have no choice but to put our trust, critically to be sure, in these basic processes, as we do in the course of everyday affairs.

Reid demonstrates that when we isolate the simplest operations of mind, there is a fundamental difference or *gap* between our immediate sensations and our perceptions of their objects, which cannot be bridged by reason, education, *or* experience; hence we must accept that we have other ways—other faculties or senses—of acquiring this information and that they are ultimately not amenable to analysis into simpler components. We are so made that we normally do not even notice that there *is* a gap, passing from the sensation to the perceived external quality without noticing that they are two disparate things. The sensation of a rose smell is not the same as an odorous effluvia (26), yet in the word "smell" we subsume both the sensation and the external power of producing it (43). "Hardness" is a quality of bodies distinct from the sensation, but we don't differentiate the sensation from the hard thing—we don't have a word for "going up against something hard"—just "hardness" that we place in the external object (55–58). If hardness is a quality distinct from our sensation of it, how do we get it? Reid argues that it cannot be by custom or reflection, so it must be "an original principle of our nature, annexed to that sensation which we have when we feel a hard body" (60).

For Reid the natural world is filled with what he calls "natural signs" that we are equipped to read, initially through innate "original perceptions" of things like the existence and primary qualities of bodies (including hardness), and the coherence and stability of nature, and then through "acquired perceptions"— which include particular laws of nature—that build on the original ones in the course of experience (58–60; 190–93). And the fundamental, bedrock enabler of higher knowledge is our intuitive apprehension of cause and effect, that is, the ongoing uniformity, connectedness, and regularity of nature, which Reid calls the "inductive principle" (198). All inductive reasoning and reasoning from analogy is grounded upon this principle, which is the basis for both science and common life, Reid positing a continuum of mental life based on the principle, stretching from "Brutes, idiots, and children" through adults to philosophers:

> All these different classes have one teacher, experience, enlightened by the inductive principle. Take away the light of this inductive principle, and experience is as blind as a mole: she may indeed feel what is present, and what immediately touches her; but she sees nothing that is either before or behind, upon the right hand or upon the left, future or past. (200)

In Reid's view, Hume wanted to deprive humanity of this innate principle,[49] and Reid's quarrel with Hume was in an important sense focused on this very point,[50] a disagreement that led him to say that "I cannot help thinking that of all the absurdities contained in Mr. Hume's system his doctrine concerning causes is chief."[51] Scholars sometimes emphasize Reid's agreement with Hume on our inability to "know" efficient causes directly—that all we can empirically observe is the constant conjunction of objects in space and time—to the exclusion of his fundamental disagreement with Hume on the nature of our perception (and hence the intelligibility) of causality.[52] On the other hand, they sometimes focus on Reid's "strict" notion of causality as active agency (and hence the subject of metaphysics rather than physics), ignoring his stress on the importance of causality in a "popular sense" for science and common life.[53] Yet Reid clearly felt that his disagreement with Hume on the roots of our perception of causality was fundamental and far-reaching; and this disagreement was based at least in part on his concern to secure and broaden the epistemological foundations of natural science.

In distinguishing his position from Hume's, Reid argues that our belief in the "continuance in the course of nature" could not be the result of a vivid association of ideas brought on by experience; such a belief is in fact an "effect of instinct" (196), it is an "instinctive prescience of the operations of nature" (198). As Reid put it elsewhere, "We are so made, that when two things are found to be conjoined in certain circumstances, we are prone to believe that they are connected by nature, and will always be found together in like circumstances." Such a belief is "an immediate effect of our constitution," it provides a basis for both natural and moral knowledge[54] and is a species of judgment: "When we attend to any change that happens in nature, judgment informs us, that there must be a cause of this change, which had power to produce it; and thus we get the notions of cause and effect, and of the relation between them."[55] In other words, causality is an irreducible principle of common sense.[56]

To give up the notion of causality as an illusion for which we have no evidence would "put an end to all philosophy, all religion, to all reasoning that would carry us beyond the objects of sense, and to all prudence in the conduct of life."[57] Reid was convinced that if we had not started with an intuitive sense of the connectedness of effects to causes, we never would have been able to get started on the road to making sense of anything. While our senses cannot give us direct information on the causal *agents* of natural phenomena, we know intuitively that there *is* a cause, and natural philosophy consists in establishing causal connections in nature, albeit at the level of rules and laws rather than agents.[58] "Gravity" was a common example: we know it by its lawful properties, not in and of itself.[59] Experience thus reveals particular causal relations in nature, but the principle of causality itself "cannot be drawn from experience any more than abstract reasoning."[60]

The common-sense assumption that we observe real causal connections in nature is not mistaken; rather, mistakes occur in the misapplication of the principle to observed phenomena, and such mistakes in thinking can be made at any level of society and education. The basic difference between correct and incorrect conclusions is that "the one is an unskilful and rash, the other a just and legitimate, interpretation of natural signs" (199–200). It is the task of science to distinguish "seeming or apparent connections of things in the course of nature, from such as are real" (200). Reid tells a story of how a very rare white ox was brought into his area, and soon afterward there occurred an uncommon fatality in childbearing, which was then attributed by the common people to the presence of the strange ox (41). This is a mis-attribution of cause and effect with no qualitative difference to mistakes made by natural philosophers—both are based on the same propensity to search after and assign causes to things, the only difference being that this one was rash and based on too little evidence. Reid thus agreed with Hume that superstition was the result of improper popular *attributions* of cause and effect.[61] The common people may indeed be mistaken in the use they make of first principles; but when it comes to ascertaining such principles, "there is no reason why the opinion of a philosopher should have any more authority than that of another man of common sense, who has been accustomed to judge in such cases."[62] First principles are what make reasoning and science possible and as such are necessarily untaught.[63] How to be sure they are correct? Valid first principles are consistent with other first principles, do not lead to absurd conclusions, and bear the consent of all nations and the learned as well as the unlearned.[64]

Does this mean truth lies in the most votes? Is common opinion a new form of authority that is going to tyrannize over mankind? Reid holds, to the contrary, that judging for oneself is an "inalienable right." But this does not preclude the fact that "Authority […] on some occasions, [may be] a useful handmaid":

> Society in judgment, of those who are esteemed fair and competent judges has effects very similar to those of civil society; it gives strength and courage to every individual; it removes that timidity which is as naturally the companion of solitary judgment, as of a solitary man in the state of nature. Let us judge for ourselves, therefore, but let us not disdain to take that aid from the authority of other competent judges, which a mathematician thinks it necessary to take in that science, which of all sciences has least to do with authority.[65]

Reid agrees with Shaftesbury (and indeed with most social theorists of the Scottish Enlightenment[66]) that we are sociable, not solitary beings, and that human beings are born for, and find the greater part of their happiness within, society.[67] And a neglected avenue of study, according to Reid, is that of "social" as opposed to solitary acts of mind. Just as modern philosophers try to reduce all of our social affections to self-love, so they reduce our social operations of mind—giving/receiving testimony, asking for/receiving information, language, promises and contracts, and so on—to solitary acts. A vast array of human knowledge and mental life—including science—occurs in interchange with other minds, yet according to Reid philosophers have proceeded as if the mind were focused primarily on solitary operations.[68] And thus it is only in private that one could doubt the reality of fundamental features of cognition—and the world—like causality, going against the wealth of available evidence while undermining the basic point of natural philosophy, which is to investigate the causes of the phenomena of nature, thereby enlarging human power, bettering the human condition, and gratifying human curiosity.[69]

Reid is clearly moving philosophy, natural philosophy particularly, out of the private realm of individual philosophers and into an emerging public sphere of universally competent actors. To the degree that all sciences rest upon first principles, their foundations lay in the widest possible public, since it is all human beings, interacting socially, who establish and entrench them. There is no sharp demarcation point between common sense and science, and in Reid's view it is incumbent upon philosophers (natural and otherwise) to act as mediators between them, to the benefit of both science and the public at large.

* * *

In grounding causality in innate features of the human mind, Reid was thus addressing not simply the erosion of the causal principle that had culminated in Humean skepticism but also the shift in natural philosophy away from its roots in common experience and toward emphasis on recondite and particular experience. Returning once again to Aristotle, the Aristotelian notion of *scientia*—causal knowledge—was concerned with revealing universal truths based on the commonly accepted facts of everyday lived experience. Premises for scientific demonstrations were taken from common life and were established as "evidence" by common consent that such and such is *what happens* in nature, like the proposition that heavy bodies fall down or fire rises up. Therefore, as Peter Dear puts it, "For Aristotle, the nature of experience depended on its embeddedness in the community; the world was construed through communal eyes." Experience provides phenomena, whose sources are diverse and include "common opinion and the assertions of philosophers as well as personal sense perception."[70] The Aristotelian model of science was thus "fundamentally public; scientific demonstration invoked necessary connections between terms formulated in principles that command universal assent."[71] Disparate particulars were no basis on which to build a science of universals, and singular and strange events and phenomena were simply "beyond the pale" of Aristotelian natural philosophy.[72] Rather, what counted for axiomatic evidence is "that which is always or that which is for the most part."[73] Moreover, it was assumed that axiomatic universal principles, including the principle of causality itself, emerged spontaneously out of sense experience and were themselves indemonstrable, it being understood that the first principles of any science must be simply taken for granted.[74] Reid's thinking represented a return to this perspective, although it had become necessary to justify it using the methods and terms of the New Science.

Earlier chapters have documented the ways in which the Aristotelian view was transformed and ultimately superseded by early-modern scientific theory and practice. Broadly speaking, over the course of the seventeenth century, the assumption that natural philosophy was ultimately rooted in the common *experience* of universal assumptions and truths gave way to the assumption that firm and useful natural knowledge grew out of close observation and description of specific, particular events and observations, including particularly *experiments*. The nature of evidence thus changed from being truths that were of common acceptance to often recondite, singular truths ("facts") that were carefully observed and attested by individuals using specialized instruments, and who established their credibility through appeal to the discursive practices of gentility, or to mathematical methodologies, or, indeed, to numerical representation itself,[75] when not assimilating singular discoveries to common experience.[76] And these "facts—observations of primarily rare or strange phenomena and experiments that probed nature's hidden secrets—were neither public nor indubitable. In contrast to Aristotelian common experience and paradigm cases they supplanted, they were unlikely on the face of it to command consensus."[77] In England they nevertheless became the foundation for new forms of civility and consensus in natural philosophy,[78] whereas on the European continent, novelties reported by those like Galileo working in the Jesuit philosophical tradition were presented within the methodological framework of common experience and were increasingly justified by recourse to mathematical methodologies and certainties.[79] By the time of Newton, the universality of natural knowledge was no longer grounded in the common experience of everyday life but in the combination of discretely observed and reported facts, often based on contrived experiments using special apparatuses and the newly respectable "mathematical" method of analysis and synthesis.[80]

The problem of induction, first raised by Hume in explicit fashion, grew out of this epistemological revolution. It had been brewing for some time with the rise of nominalism, and it came to full fruition once it was absorbed into the theory and practice of natural philosophy. Over the course of the seventeenth century, induction had been both exalted and problematized. Bacon was the first modern to really analyze it; and although he recommended it over other forms of knowledge generation, he called into question traditional understandings of what it entailed. He cautioned against premature generalization from sense experience, while elevating particulars to a foundational role in natural philosophy. The upshot was that "during the 120 years which separated the *Novum Organum* and the *Treatise of Human Nature*, very few philosophers had been entirely unaffected by doubts about the reliability of inductive inferences."[81] As long as "universal experience reigned virtually unchallenged as the irreducible touchstone of empirical adequacy," there was little problem in relating individually experienced events to universal knowledge claims about the world.[82] But once common experience no longer provided the basis for generalization, the problem became how to relate discretely experienced events to universal propositions. "When strange facts and contrived experiments replaced common experience [...] forging a shared science became problematic in principle."[83]

Not only that, but the inherited understanding of scientific knowledge itself—knowledge of causes—was threatened by the shift away from common experience. The particularizing trend of natural philosophy, in which what counted as "evident" knowledge moved from the spontaneous conclusions of common experience to the reported results of individual event-experiments, served to highlight the hidden nature of causal powers, since they did not appear to be of the kind of observable "facts" amenable to consensus that were becoming the currency of scientific endeavor.[84] Hume's philosophical skepticism toward causality thus made explicit what had been in the works for some time, and whose fate had been sealed with Newton's rejection of "hypotheses"—hypothetical physical causes—in favor of demonstrable, quantifiable behaviors, as the proper subject of natural philosophy.[85]

Thus in responding to Hume's strictures on causality, and in advancing the "inductive principle," Reid was confronting problems that had been developing within the bosom of modern science and philosophy for some time. The movement away from the agreed conclusions of common experience, like the real existence of causal connections in nature, to the inward musings of solitary philosophers, seemed to promise intellectual and social fragmentation. When viewed in this light, Reid's philosophy appears as an attempt fundamentally to rethink and reground scientific enquiry itself on more solid—and universal—intellectual and social foundations. Reid clearly was not interested in repudiating modern philosophical and methodological developments; rather, he sought to incorporate the useful insights among them[86] in a perspective that remained true to "science" as it had traditionally been understood: embedded in, rather than abstracted from, the common experience of the public at large.

* * *

That Reid's philosophy was in some measure concerned with clarifying and deepening the links between modern natural science and public culture is confirmed if we attend to the wider social and scientific contexts in which he wrote. As we have seen, by the middle of the eighteenth century a broad-based "public" with its own sphere of activity and communication had taken shape in Britain, a public sphere manifesting itself not only in urban and academic culture but also and most importantly in print, a republic of letters increasingly open to people from all walks of life.[87] These arenas of public life in Britain have been increasingly documented in the past few decades, with scholars arguing that already by the time of the English Civil War a populist political culture was coming into existence.[88] An important feature of this expanding arena of public life was a shift from traditional, courtly forms of patronage and authority to an increasingly wealthy public of middling classes based largely on the rise of commerce. Scholars have shown how the Enlightenment culture of science participated in this transition to a mass public, particularly through the popularization of science in public lectures and experiments that took place not only in London but also in provincial centers.[89] Science began playing a crucial role in helping redefine authority and legitimacy in early-modern society, in part by establishing a forceful and credible experimentalism, "but far more important[ly], by [...] creating a public science to which many might obtain entry."[90]

Whereas initially it was the inherited understandings and discursive practices of gentility, and the independence and thereby truthfulness or objectivity of the gentleperson, that provided the social basis for scientific understandings,[91] as the public sphere expanded so did the community for scientific knowledge, and all manner of popular science began to be propagated both in print and by lecturers and experimentalists. The New Science was linked to the possibility of technical and social "improvement," gaining authority thereby from its relevance to the advancement of society at large, and became a "public social badge" of enlightened civility among individuals in provincial urban centers eager to attain social status.[92] Women for their part were becoming increasingly involved in both the production and consumption of the new public scientific knowledge.[93] And scientists in Scotland began to pay attention to the relationship between their work and the public perception of it, as the presentation of science to a wider public took place within more general debates on civic life and the proper relationship between private and public good.[94]

Reid's blending of inductive natural science and commonly held understandings was right in step with this larger transformation. Toward the end of the *Inquiry* Reid states that "The science of nature dwells so near to common understanding, that we cannot discern where the latter ends and the former begins" (173). The judgments of a common man, for example, of what will or will not float, are only a few steps away from the science of hydrostatics. "All that we know of nature, or of existences, may be compared to a tree, which hath its root, trunk, and branches. In this tree of knowledge, perception is the root, common understanding is the trunk, and the sciences are the branches" (174).

Reid indicates the wider social context of his position in his opening remarks to students at the University of Glasgow, noting that knowledge of the principles of science is now required in almost all professions and activities. Scientific learning is to be pursued by all, with the goal of promoting the happiness of society, and it is science that will allow Scots to make progress in their "backward" land.[95] And it is proper to observe that

> For two hundred years back learning and knowledge has been diffused and spread among all degrees of men in Europe, and is become necessary in some degree to every man that is above the rank of day labourer. The arts have been improved and reduced to scientifick principles. The sciences have been cultivated and thrown new light upon the arts. The soldier the sailor the shipbuilder the carpenter the farmer must be men of science otherwise they cannot be eminent in their several professions.[96]

The arts and sciences were becoming linked together like never before, and consequently it was necessary for *everyone* to become conversant with science.

Some might object that Aberdeen, where Reid spent much of his adult life, wasn't exactly a hub of British public culture in the eighteenth century. But Aberdeen had a vigorous academic culture, with two colleges within walking distance of each other. And natural knowledge occupied an important position in the eighteenth-century Aberdonian academic environment, helping to "structure the public sphere by shaping popular opinion and policy."[97] In both Marischal and King's College, "scientism provided an overarching cognitive and ideological structure, linking together the moral and natural sciences in both the curricula and the clubs."[98] Aberdeen was one of a number of Scottish urban centers in which the natural sciences were increasingly a player in the formation of public culture; as such they "served as one of the defining features of public life during the Scottish Enlightenment."[99] And the Aberdonians not only were in touch with developments in the natural sciences but were deeply involved in the intellectual currents of the European Enlightenment as a whole, not least through the activities of the Aberdeen Philosophical Society.

Reid worked out his ideas in the midst of the Society, which began meeting in early 1758 and was one of many such scientific and literary societies that sprang up at the time. Reid was the driving force behind the formation of the "Wise Club," as it was called, and I would suggest that he was putting his principles into practice when he formed the club. The Wise Club was the venue in which Reid's ideas were initially made public, in a process of intersubjective validation that ostensibly kept him from the kind of speculative nonsense that he believed afflicted philosophers who spent too much time in their closets, riding their favorite hobbyhorses.[100] However unfairly it may (or may not) have been, Reid's critique of Hume and other skeptics was framed in these terms, so it only makes sense that he sought out a public forum for the development of his own ideas.

The Wise Club was devoted to the "philosophical" (inductive, observational, experimental) examination of a wide range of topics, including "every principle of Science" in a sociable atmosphere. The structure of the Club was democratic: each month a new president would assume the chair after delivering a discourse, and the group also discussed questions proposed in advance by the president and the other members. An assumption behind the posing of questions was that they would be *answered* or, in the parlance of the Club, "handled" by the group as a whole before moving on to the next question. Topics ranged from the respiration of plants, to the cause of the apparent figure of the heavens, to the nature of happiness, to the origin of civil government. The unifying factor was an inductive, "scientific" approach to the phenomena in question and a democratic, "free but friendly and candid" atmosphere of discussion.[101]

The concerns and approach of the Wise Club were clearly in tune with Reid's own philosophical perspective. What really sets Reid apart, however, is that if he was on the one hand participating in the gentlemanly, academic culture of science that had been growing since the middle of the seventeenth century,[102] he was also drawing on a wider public as a source of evidence and confirmation, thus extending the gentlemanly community of science to the public or "society" at large, not in a formal, institutional sense but in an epistemological one.[103] Reid's view of science is one that ties individual perceptions to those of a wider community, both in the search for evidence about the nature of those perceptions and for intersubjective confirmation of them. As with many of his predecessors in the common-sense tradition, Reid often appeals to the perceptions shared by all persons in society, and/or in all times and places, to verify his claims, and this might be troubling if Reid didn't distinguish between fundamental common-sense apprehensions and more complex ones, based on experiment and reasoned analysis, and if he didn't provide a coherent account of how error occurs on the principles of his system.[104]

As we have seen, early in the eighteenth century the Third Earl of Shaftesbury had propounded an influential moral philosophy that posed the "sensus communis" or "common sense" of the "the publick" against the skeptical and selfish conclusions of "solitary" modern philosophers.[105] Working in the legacy of Shaftesbury, Francis Hutcheson similarly answered modern "selfish systems" of philosophy with an analysis of our benevolent and sociable impulses as expressed in, and validated by, the "public" at large.[106] It was not a large step to transpose this discourse from morals to epistemology and natural knowledge, and Reid—with the help of Turnbull and Kames—did just that. Reid's own political philosophy expresses a very strong, Ciceronian emphasis on the *res publica* as a central (although not the only) touchstone for individual behavior and motivation, which when taken to the extreme of speculating about a "utopian system" resulted in Reid arguing for a polity without private property and in which "no private interests [would be] opposed to that of the Publick."[107] Reid's emphasis on the common or public good provided a framework for individual rights and duties, as he blended natural jurisprudence with humanist optimism in his political theory.[108] And Reid's initial support for the Revolution in France, lasting until around 1794 and evidenced by a donation to the National Assembly, attests to his republican convictions.[109]

Reid believed that moral values like justice, prudence, and regard to the common or public good were universally held in all times and places and were in the nature of things and that human beings had an intuitive understanding of them, an understanding that they were free to act upon in the public sphere and that such freely willed, rational moral actions were the building blocks of social order.[110] Hume, on the other hand, argued that morals emerged out of a combination of self-interest, a sympathetic identification with the feelings of others, and feelings of pleasure and pain, and thus although in some sense "natural" were "perceptions in the mind" "more properly felt than judged of."[111] He envisioned a modern, commercialized public sphere inhabited by passionate, propertied, self-interested actors whose interests needed to be appeased and whose opinions enlightened by philosophers able to show them that intuitive perceptions of the public good were often mistaken.[112] Hume may have entertained his own naturalistic brand of moral realism,[113] and he believed there to be a "middling rank of men, who are the best and firmest basis of public liberty,"[114] but he did not place the same trust in the opinions of ordinary people that Reid did,[115] although it should be added that as an educator Reid was well aware of the need for education to produce both good science and good citizens.[116]

Reid's philosophical perspective balances intuitive, individual, and private perceptions with common, public perceptions, aiming to do justice to both. If it seems at times that an *argumentum ad populum* is being advanced, it is underwritten by close attention to individual human perceptions and an emphasis on the necessity of independent judgment.[117] "Common sense" is the generalization that results from the plurality of human experience, the distillation of the essential principles of human nature. All sciences rest on these first principles, grounded on evidence corroborated by the widest of all possible publics.

James Beattie, Reid's disciple and popularizer, noted in 1773 that the foundational importance of common-sense notions was "no modern discovery."

> Aristotle treats of self-evident principles in many parts of his works [...] and says of them [...] That they are known by their own evidence; that except some first principles first be taken for granted, there can be neither reason nor reasoning [...] that if ever men attempt to prove a first, it is because they are ignorant of the nature of proof.[118]

Reid for his part noted in the *Inquiry* that "The old [Peripatetic] system admitted all the principles of common sense as first principles, without requiring any proof of them; and therefore, though its reasoning was commonly vague, analogical, and dark, yet it was built upon a broad foundation, and had no tendency to scepticism."[119]

In his essay on Aristotle's logic, first published in 1774, Reid wrote that one of Aristotle's "capital conclusions [...] of which his genius was wonderfully productive" was that "All demonstration must be built upon principles already known; and these upon other of the same kind; until we come at last to first principles, which neither can be demonstrated, nor need to be, being evident in themselves."[120] Later in the essay, Reid concludes that

> Men rarely leave one extreme without running into the contrary. It is no wonder, therefore, that the excessive admiration of Aristotle, which continued for so many ages, should end in undue contempt; and that the high esteem of logic as the grand engine of science, should at last make way for too unfavourable an opinion, which seems now prevalent of its being unworthy of a place in a liberal education.[121]

Thus despite the weaknesses and excesses of the peripatetic philosophy (including a "redundance" of first principles[122]), Reid clearly believed that Aristotle had gotten some very fundamental things right.

Finally, in his *Essays on the Intellectual Powers of Man*, Reid states that Buffier was "the first, as far as I know, after Aristotle, who has given the world a just treatise upon first principles,"[123] again indicating the continuity he saw between his own philosophy and that of the tradition of thinking we have been following in this book. Thus after many years of often virulent attacks on Aristotle, an ardent follower of Bacon and Newton recognized the cogency and utility of Aristotelian foundationalism, albeit reformulated in terms suited to the social and scientific requirements of modernity. As a natural philosopher and mathematician who could read and understand Newton's *Principia* from cover to cover, Thomas Reid was in a good position to be magnanimous toward the common people, while challenging a philosophical position that had become both cognitively and socially problematic. And his *rapprochement* between science and common sense came at the problem from both ends: Reid was concerned that "philosophy" would not be alienating to the common people, while at the same time demonstrating to philosophers that all human beings come equipped with the basic aptitudes to participate in science and that common intellectual processes differ only in degree, not in kind, from those of science. If Newton's rules of reasoning in philosophy were truly "maxims of common sense," then everyone, in some small way, could be counted a Newtonian. Reid's philosophy thus brought the long-running discussion about the relationship between common sense and science full circle. Aristotle would doubtless approve.

EPILOGUE

Thomas Reid's philosophy of common sense exercised a significant influence on Western thought and continues to be relevant today. It appeared on the world stage as a scientific philosophy amenable to Christian beliefs, the rise of a modern public sphere and democratic politics. Exercising its most profound impact in postrevolutionary France and America, it promised to combine progress and stability, establishing links between common-sense experience and philosophical and scientific thought in an era of rapid sociopolitical, religious, and scientific change. It had a significant impact on the development of higher education in both countries and was an important undercurrent in the broad expanse of nineteenth-century intellectual culture, a current which fed and mingled with other streams of thought.

Although an identifiable school of "common-sense philosophy" began to wane around the middle of the nineteenth century, Reid's philosophy proved to be a multivalent and fertile influence on subsequent philosophical developments in Britain, France, and America. Reid's impact in German-speaking lands was slight but worth considering. And today his thinking remains pertinent not only to philosophers but also to psychologists, cognitive scientists, and philosophers of science. In what follows, I first provide an analytical overview of Reid's historical influence and impact in Britain, France, Germany, and America, before concluding with some observations on the relevance of his thought to the findings of contemporary research into the nature and structure of the human mind and to the ongoing scholarly discussion about the relationship between common sense and scientific inquiry.

* * *

Reid's thought gained an immediate following upon publication of his *Inquiry* in 1764. James Oswald (1703–1793) was a Scottish minister who published *An Appeal to Common Sense in Behalf of Religion* in two volumes (1766/1772). Although he cited Reid sparingly, Oswald seized upon Reid's ideas to defend the faith from "the assertions of sceptics and infidels,"[1] opposing the common-sense principles that ordinary people take for granted to the labored and drawn-out reasonings of the learned, the latter threatening to confound the instinctive moral and religious impulses possessed by all human beings.

A somewhat more substantial defense of religion and traditional morality soon followed from the pen of James Beattie (1735–1803), appointed Professor of Moral Philosophy and Logic at Marischal College in 1760. Beattie joined the Aberdeen Philosophical Society in that year and thus came into personal contact with Reid during the formative stage of Reid's *Inquiry*. In his *Essay on the Nature and Immutability of Truth*, first published in 1770, Beattie mounted a vociferous attack on Hume, using arguments drawn from Reid but with little of Reid's characteristic subtlety or depth. In Beattie's view, nothing less than the existence of truth itself was at stake in the debate with Hume and other skeptics and infidels. Religion, morality, and civilized behavior were teetering on the edge of the abyss, and it was time to rise up and put a stop to the nonsense before all was lost. Echoing Reid, Beattie emphasized the public nature of all forms of higher knowledge and their roots in the common sense of humankind. Just as mathematical reasoning falls to the ground if one doubts the axioms of geometry, so also if one supposes that "the dictates of common sense are erroneous or deceitful, all science, truth, and virtue are vain."[2]

Beattie combined a dire sense of imminent cultural collapse with enough of an understanding of Reid's ideas to produce what amounted to a literary blockbuster. Initial reviews of the book were favorable, the work being praised by Samuel Johnson, Thomas Percy, David Garrick, and Edmund Burke, among others. It even merited praise from King George III, who granted Beattie a royal pension of £200, while Joshua Reynolds painted a portrait of Beattie with his *Essay* under his arm and the Angel of Truth hovering nearby. The *Essay* became an important conduit through which Reidian thought was made known outside of Scotland, going through 14 English editions by the end of the century, and translated into German, French, and Dutch.[3]

Reid attained deeper and more lasting influence through another disciple, Dugald Stewart (1753–1828). Entering the University of Edinburgh in 1765, Stewart was inspired by Newtonian physics and Baconian inductive method. He attended Reid's lectures at Glasgow in 1771–72, and Reid's impact on Stewart is evident in nearly all of Stewart's writings. In 1785 Stewart was appointed to the Chair of Moral Philosophy at Edinburgh, a position he held until his retirement in 1810. His works included *Elements of the Philosophy of the Human Mind* in three volumes (1792, 1814, 1827), *Outlines of Moral Philosophy* (1793), and *Philosophical Essays* (1810). These works (and especially the *Elements*, which were extensively reprinted in France and America) sought to develop the scientific and practical applications of Reid's thought in a broad social and intellectual context.

In the first volume of the *Elements*, Stewart sought to demonstrate how an inductive science of the mind is relevant to improvements in education, science, politics, and the arts. For example, Stewart asserts that education will be improved by clarifying the distinctions between basic features of the mind and acquired associations and prejudices. Science will benefit from a clearer understanding of the relationship between the principles of our nature and the phenomena of physical nature, an understanding that will discourage unwarranted hypotheses and encourage a focus on verifiable facts and laws of nature. Political and economic thought must be grounded in common experience and understandings, if it is going to be useful. The arts similarly gain by an understanding of the formative role played by basic human propensities (including the power of association) in poetry, painting, eloquence, and other fine arts.[4]

In his discussion of modern social and political developments, Stewart looks with favor on the rise of the commons and public opinion, chiding the French monarchy for having ignored them for so long that they finally exploded in its face.[5] In Stewart's discourse, the common people, endowed as they are with what he would later call "fundamental laws of belief," appear as the foundational units of the whole social structure, yet they must be educated and enlightened by the right sort of scientific philosophy if they are to be put on an appropriate path. Here we see an agenda for healing the rift between common sense and science, emerging directly out of the common-sense philosophical tradition (and Reid's thought in particular) that is relevant more than ever today. The question is one of emphasis rather than substance; if we can truly grasp the deep epistemological and social connections between common sense and science, we will be motivated to redouble our efforts (and spending) on science education for everyone. By the same token, if science is understood to be rooted in common sense, it becomes more salient *politically* to emerge as a champion of both "the people" and science/science education.

Stewart's second volume, appearing more than 20 years after the first, is taken up with more strictly philosophical topics. Stewart uses the example of mathematics to critique Reid's language of "principles of common sense," arguing that what Reid was talking about were neither "principles" nor "common sense," as both are normally understood. Voicing what was to become a common criticism, Stewart argued that the term "common sense" is problematical because it is loose and ambiguous and implies "common opinion" or even "prejudice."[6] Stewart's solution was to drop the term entirely and urge the more respectable, and less ambiguous, "fundamental laws of human belief" on his readers. Thus if the first volume of the *Elements* argued for the utility of Reid's philosophy, the second volume sought to assure its intellectual respectability.

Stewart was influential not only through his writings but also through his students, many of whom went on to become politicians, religious leaders, and men of letters, including Sir Walter Scott, Thomas Chalmers, James Mill, and Thomas Brown.[7] Brown assumed Stewart's chair in moral philosophy from 1810 until Brown's death in 1820. A medical doctor, Brown continued to follow the "scientific," psychological method of his predecessors (adding his own physiological elements), and he argued strenuously for the existence of intuitive beliefs or principles of the mind. Brown's lectures were clearly indebted to Reid, and he often cited Reid's works, but he also argued that Reid's critique of "the theory of ideas" was misguided, and he criticized Reid's account of perception.[8] In his *Inquiry into the Relation of Cause and Effect* (1818) Brown criticized Reid's identification of "power" with agency, arguing that a relation of uniform antecedence and consequence is all that terms like "power" and "cause" really mean. At the same time he argued that our belief in such notions is intuitive rather than derived from custom or experience. Brown, who was widely read in America, thus staked out something of a middle position between Reid and Hume on the issue of causality.[9]

It should not be surprising that as time passed, Reid's ideas were subjected to critical scrutiny, amended, and joined to other ideas and doctrines, while remaining a central undercurrent of nineteenth-century British thought. One figure who took Reidian thought in a new direction was William Hamilton (1788–1856), who became Professor of Logic and Metaphysics at Edinburgh in 1836 and was a towering figure in British philosophy until his death in 1856. Thomas Carlyle, Clerk Maxwell, and James Lorimer were among the members of Hamilton's circle, "a seed-bed of cultural resurgence" in the 1830s and 1840s.[10] It was through Hamilton's heavily annotated and oft-reprinted edition of Reid's works that many in Britain and America were exposed to Reid's philosophy in the second half of the nineteenth century.

Hamilton, like a number of others at the time, assimilated Reid's ideas to Kantian philosophy, and he used Reidian notions, including the existence of first principles of mind and the need for critical analysis of the contents of consciousness, as a starting point for his own metaphysical inquiries. His "philosophy of the unconditioned" led to a kind of agnosticism and phenomenalism in the eyes of some,[11] while others saw it "as a mean between the one extreme of a thoroughgoing scepticism and the other extreme of the monistic omniscience of Gnosticism."[12]

James Ferrier (1808–1864) was a friend and admirer of Hamilton who, after an unsuccessful bid to assume Hamilton's Chair at Edinburgh, became Professor of Moral Philosophy at St. Andrews. Ferrier's thought was described by contemporaries as "German philosophy refracted through a Scottish medium"[13] and as such signaled the end of Reidian hegemony and the turn toward various forms of idealism in British universities during the second half of the nineteenth century. Ferrier was highly critical of Reid's thought, arguing that Reid had accepted premises central to the ideal theory, namely the existence of an object (the external world) separate from a subject (the perceiving mind) which represents that world to itself. Ferrier believed that this distinction between inquiring subject and observed object is violated in the very act of introspective analysis, and as such constituted a "radical defect" in Reid's psychological method and the science of the human mind in general. But whereas Brown clearly had Reid's texts in front of him while leveling his criticisms, Ferrier seems to have been responding more to a vague (and rather distorted) memory of Reid's ideas, to Brown's reductive, "scientific" approach to mental phenomena, and to the growing hegemony of natural science in general.[14]

Thus by the middle of the nineteenth century Reid's thought had been critiqued from both reductionist (Brown) and idealist (Ferrier) perspectives, while Hamilton had effected an idiosyncratic adaptation of Reid's thought to Kantianism. Hamilton's thought was itself subjected to devastating criticism by J. S. Mill in 1865, an attack that contributed to the decline of the commonsense tradition, by then strongly associated with Hamilton, in Britain as well as in America.[15] Mill's frontal assault on Hamilton can be seen as one episode in a long struggle between an optimistic scientism that held that scientific specialization and technical education were the keys to the future, and the Reidian notion, as stated by George Davie, "that the scientific and technical expertise required by modern civilization will turn into an unintelligible and lifeless routine if it is allowed to develop in a departmentalised way, out of touch with the common sense of the lay populace."[16] This observation clearly retains its relevance today.

Just as the Reidian thread was becoming attenuated in Scotland, in England it was being woven into the thought of Henry Sidgwick and his student G. E. Moore. Sidgwick (1838–1900) was probably exposed to Reid as a student at Cambridge under William Whewell. Sidgwick's early writings on epistemology "suggest a common-sense approach to problems he was later happy to associate with Reid," and his own ethical position "incorporates much of the substance of Reid's [ethical position]."[17] That Sidgwick felt strong affinities for Reid was made clear in an 1895 lecture delivered to the Glasgow Philosophical Society titled "The Philosophy of Common Sense." In the lecture, Sidgwick rebutted Kant's critique of Reid (discussed below) and praised Reid's psychological method and his separation of sensation and perception. And Sidgwick "[could] not think Reid wrong in holding that the propositions he is most concerned to maintain as first principles are implicitly assented to by men in general." This was not to say that such beliefs are always correct.

Reid's "essential demand [...] on the philosopher, is not primarily that he should make his beliefs consistent with those of the vulgar, but that he should make them consistent with his own."[18]

George Edward Moore (1873–1958) studied under Sidgwick at Cambridge and, in giving his own stamp to Reidian thought, influenced a generation of British intellectuals.[19] Moore is widely seen to have spearheaded the attack on idealism and to have turned British philosophy away from questions about whether ordinary claims concerning the external world and morality have truth or meaning, toward the analysis of such claims.[20] Despite the fact that Moore rarely referred to Reid, it is clear that he had studied Reid's works, writing approvingly of Reid on occasion, and the connections and affinities between the two have been noted by scholars.[21] In "A Defense of Common Sense" Moore employed his careful analytical method to defend the truth of common-sense convictions such as "There exists at present a living body, which is *my* body,"[22] arguing that denials of such truisms are either self-contradictory or factually false. Moore was thus led to say that "I am one of those philosophers who have held that the 'Common Sense view of the world' is, in certain fundamental features, *wholly* true."[23] Thus despite going out of fashion during the latter decades of the nineteenth century in Britain, Reid's thought played an important role in the rise of the analytic movement in the twentieth century.

* * *

Although German philosophers became aware of Reidian thought early on, it did not have much of an impact on the mainstream of German philosophy. The paradox of the "German reception" of Reid is that most sympathetic Germans did not accept two of the essential elements of Reid's thought: the rejection of the theory of ideas and the irreducibility of principles of common sense.[24] If Reid, Beattie, and Oswald helped to focus German interest on first principles of mind and the role of sense experience in cognition, they were also widely seen to be advocating a perspective antithetical to philosophy itself, a "misology, reduced to principles" in Kant's words.[25] The rationalistic tradition of German philosophy, coupled with the absence of sociopolitical conditions that made recourse to common-sense understandings attractive,[26] contributed to the generally tepid response to Reid's thought in the German states.

There were a number of avenues by which Germans could become aware of the works of Reid, Beattie, and Oswald. French and German journals reviewed Reid's *Inquiry*, which was translated into French in 1768 and into German in 1782. Oswald's *Appeal* and Beattie's *Essay* were widely reviewed in German periodicals, and Beattie's *Essay* was quickly assimilated after being translated into German in 1772.[27] Reid's thought was thus primarily made known through the writings of Beattie and Oswald, and Joseph Priestley's *Examination* of common-sense philosophy, published in 1774 and reviewed in the *Göttingsche Anzeigen*, helped to fix the notion in German minds of an inter-changeable triad "Reid, Oswald, and Beattie." In that work, Priestley argued that Reidian thought puts up barriers to scientific research into mental phe-nomena, ascribing too much importance to intuition and instinct, and too little to reason and education.[28]

Yet Reidian thought did spark some interest in German lands. Beattie's work had an influence on the Göttingen thinkers Christoph Meiners and Georg Lichtenberg, while the writings of Johann Feder (1740–1821) exhibit the impact of both Beattie and Reid, Feder warmly praising both Reid's *Inquiry* and *Intellectual Powers*.[29] In *Logik und Metaphysik* (1770), Feder follows Reid and Beattie in arguing for the existence of "first principles that reasoning does not clarify and illuminate, rather confuses." And in his discussion of "The sources of truth and the grounds of the reliability of perception," he argues for the reliability of sense perception in Reidian terms and cites Reid and Beattie for further reading.[30] *Logik und Metaphysik* thus contains one of the most straight-forward presentations of Reidian ideas in German letters, yet as a whole it is an eclectic and synthetic work, Feder citing many modern philosophers—Reid only rarely, while Hume is cited more often than Reid, and Locke more often than either of them.[31] Feder furthermore did not reject the theory of ideas, and "[He] never became a follower of the Scots in the sense of accepting all or even most of Reid's theory of knowledge."[32]

Johann August Eberhard (1739–1809) propounded a Reidian distinc-tion between sensation and perception, but he accepted only a few very basic principles of common sense such as the principle of noncontradic-tion, and he felt, as most German thinkers did, that the Reidian insis-tence that such principles were not open to rational scrutiny represented a turning away from scientific and philosophical analysis.[33] Eberhard's notion that principles of common sense required clarification and rectifica-tion by rational analysis was further developed by Johann Nicolaus Tetens (1736–1807), whom Kuehn calls "the German philosopher most influenced by the Scots."[34] But Tetens argued that Reid's response to Hume was "not wrong, only unphilosophical"—not much different than if a natural sci-entist were to explain the phenomenon of magnetism by saying that "the magnet drew the iron to it by instinct."[35]

There is not much evidence that Reid exercised a significant influence on Immanuel Kant (1724–1804). If Kant's dismissive attack on Reid and his epigones in the *Prolegomena to Any Future Metaphysics* (1783) indicates that Scottish common-sense philosophy was a part of the intellectual climate in which Kant's works were formulated,[36] it also demonstrates that Kant most likely had not studied Reid and that instead he was basing his understanding of Reid on Priestley's *Examination*, or perhaps on Oswald's *Appeal* or Beattie's *Essay*.[37] Whatever the case, if one wants to argue that Kant read Reid in any depth, then one must also assert that Kant either did not understand him or that he willfully misrepresented him, arguing as he did that Reid missed the point of Hume's problem and that "Seen clearly, this appeal [to common sense] is nothing but a call upon the judgment of the multitude, whose applause embarrasses the philosopher, while the popular wiseacre glories and boasts in it."[38] In his discussion of common-sense philosophy, Kant uses four different German expressions in referring to common sense, none of which approximates Reid's own more precise notion of "principles" of common sense, further throwing into doubt Kant's familiarity with Reid's thought.[39]

Two of Kant's principal philosophical opponents, J. G. Herder and Friedrich Nicolai, freely and uncritically appealed to common sense in their works, albeit without reference to Reid. Both thinkers ascribed to a populist conception of Enlightenment, and Herder, in contrast to Kant, grounded his philosophical perspective in the understandings, language, and traditions of the German *Volk*.[40] F. H. Jacobi, another critic of Kantian thought, appealed to a primordial certainty about what is real "that animates our experiences from the beginning and pervades all levels of language." Jacobi called this certainty "faith," and in clarifying what he meant by it made reference to Reid's *Intellectual Powers*, although it is unclear whether he had actually read Reid's works himself.[41] J. G. Hamann, another swimmer against the German metaphysical current, appears to have read Reid and espoused ideas of natural language and original perceptions similar to Reid's.[42]

Kant's cutting remarks and lack of familiarity with Reid's philosophy highlight the overall lack of affinity between Reidian thought and the more rationalistic mainstream of German metaphysics. In his *Lectures on the History of Philosophy*, G. W. F. Hegel exhibited a better understanding of Reid than did Kant, yet it appears that he too was not directly familiar with Reid's work.[43] It is more than a little ironic that over the course of the nineteenth century many individuals came to see Reid and Kant as having been engaged in similar enterprises (with some fundamental differences between them). Yet it is hard not to sense, in Kant's vehement attack on Reidian thought, a defensive reaction to the underlying suggestion of the common-sense school that rational analysis of the a priori conditions for the possibility of knowledge— Reid's irreducible principles of common sense—was misguided.

* * *

Reid had a major impact on French thought in the first half of the nineteenth century, particularly on the so-called "Eclectics" (or "Spiritualists") who forged a position between conservatives horrified by the French Revolution and radicals who felt that its aims had not been fully realized. Although Reid's *Inquiry* had been translated into French in 1768, it was not until the first decades of the nineteenth century that Reid's thought aroused significant interest, as some individuals—including Traditionalists like de Maistre and de Bonald— began calling into question the legacy of the French Enlightenment, including its more materialist currents. Maine de Biran (1766–1824), who read and praised Reid,[44] developed an influential notion of metaphysics "as the science of principles [...] found in the primitive facts or basic data of intuition."[45] The way had been cleared as early as 1804, when Joseph Marie de Gérando included a section on "the Scottish School" in his *Histoire comparée des systémes de philosophie*, giving a mixed review of Reid while reserving heartier praise for Stewart, whose *Elements of Philosophy* appeared a few years later (1808) in French translation.[46]

The story of Reid's reception in France usually begins, however, with Paul Royer-Collard (1763–1845), a political figure appointed by Napoleon in 1811 to the chair of philosophy at the Sorbonne. It was through Royer-Collard's lectures there that Reid became widely known in France. In his opening lectures to the course in the History of Philosophy, as well as in his other literary remains, Royer-Collard recapitulated some of Reid's central insights into the nature of perception and the philosophical basis for scientific knowledge of the world. He reiterated Reid's critique of modern philosophy and gave Reid credit for "utterly destroying" the theory of ideas.[47] Royer-Collard furthermore provided a detailed Reidian answer to Hume's problem of induction, arguing that we are enabled to grasp causal relations in nature by the "primitive facts" of consciousness and memory.[48] "When one goes against [such] primitive facts, one misunderstands both the constitution of our intelligence and the goal of philosophy," which, like the other sciences, is to attend to facts and avoid hypotheses. The result of the interaction of sense, consciousness, and memory is that "order shines on the universe and man learns to read in the great book of nature."[49]

Generally speaking, Royer-Collard employed Reid's thought to affirm our ability to have knowledge of ourselves, the physical world, and of "a first and necessary cause [...] whose power and extension is equal to the magnificence and harmony of the effects it produces before our eyes."[50] Royer-Collard, and those who followed in his footsteps, thus found a way around the tradition of French philosophy stretching from Descartes through Condillac that, in Royer-Collard's view, was overly hypothetical and had led to a skepticism and materialism with destructive consequences. "When all existence is in question, what authority remains in the relationships that unite it? Yet on these relationships depend all the laws of societies, all the rights, all the obligations that constitute public and private morality."[51]

When Royer-Collard stepped down from his teaching post in 1814, Victor Cousin (1792–1867), a student of Royer-Collard's, took over his lectures in the history of philosophy, seeing himself to be "continuing the investigations of our illustrious predecessor."[52] *Du vrai, du beau et du bien* (*The True, the Beautiful, and the Good*; 1836/1853) was drawn from course lectures delivered starting in 1817 and contains representative doctrines that Cousin reaffirmed late in life.[53] In the book, Cousin builds an eclectic philosophy on foundations laid by Reid and to a lesser degree Kant. Highest praise is reserved for Reid, of whom Cousin declares himself a disciple: "We regard Reid as common sense itself, and [...] common sense is to us the only legitimate point of departure, and the constant and inviolable rule of science. Reid never errs; his method is true, [and] his general principles are incontestable."[54]

Cousin argues at length that "universal and necessary principles" of mind such as space, time, and causality "are encountered in the most common experience"[55] and undergird all the sciences, while also leading to the perception of ideal truths and the existence of God. As such, Cousin presents his "philosophical science" as a form of "idealism rightly tempered by empiricism"—that is, a reconciliation of sense experience with absolute ideas (in the sense of Platonic Forms independent of experience and consciousness) that are revealed by reason.[56] Absolute "ideas" or "truths" of science, art (beauty), and ethics (goodness) ultimately find their source in God but are revealed in experience by reason, a spontaneous faculty that "discovers" universal and necessary principles of common sense without "passing through analysis, abstraction, and deduction."[57] "Spiritualism" was Cousin's term for this doctrine, whose "character in fact is that of subordinating the senses to spirit, and tending, by all the means that reason acknowledges, to elevate and ennoble man."[58]

Cousin achieved public fame in the late teens and twenties as an innovator in philosophy and a champion of liberty who had spent time in a Prussian prison for his views. Once restored to his teaching post at the École Normale in 1828 (it was closed in 1820), his lectures were reportedly printed and sold to the tune of 3,000 copies.[59] In the decade after the accession of Louis-Phillipe in 1830, Cousin became a dominant figure in French political and educational life, becoming a member of the Superior Council of Public Instruction, president of the National Teacher's Examination in Philosophy, director of the École Normale, minister of Public Instruction, and president of the Academy of Moral and Political Sciences, finally retiring from public life amid the tumult of 1848.

In these various capacities Cousin played a major role in the institu-tionalization of philosophy instruction at secondary and university levels, establishing "Eclecticism" or "University Philosophy" or "New French Philosophy" (all synonyms) as the standard philosophy curriculum of higher education. According to Patrice Vermeren, "Cousin's sweep into hegemony in the 1830s and 1840s represents the first time philosophers became men of state in France, and his model for the place and role of institutions of knowl-edge would determine their course ever since."[60] If Cousin imposed a philo-sophical orthodoxy on French institutions of higher learning, he was also an intermediary and catalyst in the transition from an elitist social system to a modern merit-based society. Through the institutionalization of philosophy, Cousin helped to make the university, rather than the salon, the focal point and training ground of liberal society.[61]

Cousin's many followers were also for the most part followers of Reid. Most prominent among them was Théodore Jouffroy (1796–1842). Jouffroy had studied under Cousin at the École Normale, and during the 1820s he hosted an influential salon in his own apartments, before being appointed Professor of Ancient Philosophy at the Collège de France in 1833. Jouffroy's writings included philosophical essays and courses on natural law and aesthetics, and he translated Dugald Stewart's *Outlines of Moral Philosophy* and the complete works of Reid, attaching lengthy, influential prefaces to both works outlining his own views.

Jouffroy was concerned with establishing "philosophical science" on a par with the physical sciences, in terms of both establishing fundamental facts and laws of the mind and achieving public stature and credibility comparable to that of natural scientists. If Cousin had paid lip service to science, Jouffroy, who for the most part rejected Cousin's spiritualism,[62] intended to flesh out the potentialities of Reid's thought for putting philosophy onto a firmly scientific, yet antireductionist, footing.[63] For example, in Jouffroy's view the common-sense judgment that mind and body are of a fundamentally different nature provides a point of departure for psychological research, underwriting the notion that the object of psychology is empirical study of the self-conscious life force or soul.[64]

For Jouffroy, then, Reidian thought placed philosophy onto scientific footings, helping to lift it out of its "impotent" state and paving the way for philosophical science to assume a public stature on par with the natural sciences.[65] And there can be little doubt that Royer-Collard, Cousin, and Jouffroy became arbiters of public opinion to a degree that philosophers rarely attain, common-sense philosophy receiving acceptance by a large part of society and occupying a dominant position in higher education in France until it began a slow decline around 1870.[66] Reidian thought also influenced the aesthetic theories of

Cousin, Jouffroy, and those who followed in their footsteps, including Adolphe Garnier, Charles Lévêque, René Sully-Prudhomme, and C. A. Sainte-Beuve.[67]

The extent of Reid's impact on nineteenth-century French intellectual life is indicated by the fact that the positivism of August Comte, which was gathering adherents by the middle of the century, was firmly grounded in what Comte variously called "universal good sense," "vulgar wisdom," "common reason," and "simple good sense." According to Comte, "Science, properly speaking, is simply a methodical extension of universal good sense. Far, therefore, from treating as questionable what has been truly decided by it, healthy philosophic speculation must always be indebted to common reason."[68] Comte, who lectured in Paris during the years of Cousinian hegemony, advanced a view of the relationship between common sense and science clearly in tune with that of the common-sense school. He characterized eclecticism as a useful if rather impotent "stationary" school of thought, and positivism as the way forward in the ongoing effort to reconcile order with progress.[69] This had been the attraction of Reidian thought in France in the first place, providing, according to Charles Rémusat, a scientific foundation on which to rebuild philosophy and social order in an age when "one only hears about [...] the uncertainty of theories, the vices of institutions, the instability of governments, the decadence of the arts and letters, the lowering of character, the rarity of talent, the weakness of mores, the loss of convictions, [and] the dangers of industry."[70]

* * *

Translations of works by Cousin and Jouffroy were published in the United States starting in the early 1830s, contributing to the pervasive influence of Reidian philosophy in nineteenth-century America. Reid's thought played an important role in the conservative evangelicalism of Princeton and the American South, the Unitarianism and transcendentalism of the New England states, and the institutionalization of science and science education in universities throughout the country. A profusion of textbooks ensured that Reidian thought, in the guise of mental and moral science, spread far and wide, while at the end of the century it was assimilated into the pragmaticism of C. S. Peirce. Common-sense philosophy was thus one of the primary conduits by which the perspectives of the New Science and the Enlightenment were received in America, and it had an impact on many subsequent developments in American philosophy and higher thought.[71]

Reid's influence on American thought was already evident in the eighteenth century. Thomas Jefferson was familiar with Reid's *Inquiry*, and a number of Jefferson's key ideas, including the notion of "self-evident truths" as employed in the Declaration of Independence, can be traced to the influence of Reid.[72]

Another founding father, the visionary legal scholar James Wilson, was deeply influenced by Reid. Wilson gave a celebrated series of lectures at the College of Philadelphia in 1790–91 in which he criticized Locke and praised Reid, maintaining that while the skepticism of Hume was subversive of liberty and responsibility, Reid's philosophy of common sense offered a scientific confirmation of an innate moral sense that could be relied upon to serve as a secure egalitarian basis for law and politics in the new republic.[73]

Another early American proponent of Reidian thought was John Witherspoon (1723–1794), a respected minister in the Church of Scotland who became president of the College of New Jersey in 1768. In America Witherspoon became a prominent Presbyterian leader and supporter of American independence, serving in the Continental Congress from 1776 to 1782. At Princeton he broadened the curriculum on the pattern of the Scottish universities, encouraging the study of natural philosophy and introducing many of the leading lights of the Scottish Enlightenment to his students. Under Witherspoon's watch the College attained national stature as a training ground for public leaders.[74]

As was the norm for American college presidents at the time, Witherspoon taught the course in moral philosophy. In his lectures he revealed himself to be an eclectic follower of Hutcheson and other British moralists, concerned to harmonize the claims of revelation, reason, and conscience. Witherspoon introduced Reid and Beattie to students and propounded a philosophical response to idealism and skepticism that was broadly similar to Reid's philosophy, thus helping prepare the way for the American appropriation of Reid's thought.[75] Many of Witherspoon's students rose to prominence, assisting the spread of the Scottish Enlightenment—and Reid's ideas—in America. They included over 100 ministers, 13 college presidents and many more college educators, a US president (Madison), a US vice president (Burr), and 20 US Senators.[76]

It was under Witherspoon's successor and son-in-law Samuel Stanhope Smith (1750–1819) that the College came fully under the influence of Reid and the philosophy of common sense. Smith developed the more liberal and progressive intellectual tendencies of Witherspoon and worked to expand and deepen the natural science curriculum at the College, purchasing scientific instruments and appointing the first professor of chemistry in America, John Maclean, in 1795.[77] Smith's wide-ranging *Lectures* [...] *On Moral and Political Philosophy* (1812) was the first major exposition and application of Reid's thought written and published in America. Based on class lectures spanning the previous 30 years, the two volumes indicate that Smith had adopted Reid's perspective and terminology in forging a *rapprochement* between science and religion. Modern Newtonian science had banished hypotheses and

concentrated on the formulation of general rules of nature's effects by inductive analysis; similarly, through inductive analysis of the powers of mind, as
well as from external evidence, we can discern the laws of our constitution and
the intentions of our creator. Smith recapitulates Reid's critique of the theory
of ideas as an unsound hypothesis, preferring to "form judgments [based] on
experience and fact, interpreted by plain common sense," and he lauds Reid
for having placed philosophy on a proper foundation, "that common sense
which it had deserted."[78]

The *Lectures* apply a Reidian perspective on everything from sense perception
to volition to language (where Reid's view on natural language is presented),
to natural theology, duty, aesthetics, economics, and politics. Smith's scientific
proclivities are evident throughout: the lecture on internal sense perception,
for example, culminates in a discussion of Hartley's theory of "vibratiuncles"
as a way to account for madness and nervous diseases (Lecture VII). When
it came to natural theology, Smith rehearsed a Reidian critique of Hume's
position on causality, asserting that an instinctive inductive principle is a basic
feature of our constitution that makes all forms of knowledge possible and
leads us to infer the existence of God from the effects we see in the world
(Lecture XV).

Smith was widely seen in his day as a progressive educator and elo-
quent speaker, and many of his students went on to assume leading roles in
American churches and universities.[79] He and Witherspoon were thus in many
respects responsible for bringing the philosophy of Reid into wide circulation
in America, including in the American South.[80] But Princeton proved not to
be conducive to Smith's kind of liberalism, and he resigned under pressure in
1812, ushering in a more conservative, evangelical era at the College, led by
Archibald Alexander and Charles Hodge, who employed common-sense philosophy for primarily religious and didactic ends.[81]

James McCosh (1811–1894) became president of the College in 1868.
Before coming to America, McCosh had published *The Intuitions of the Mind,
Inductively Investigated* (1860), in which he sought to clarify and delimit the nature
and mode of operation of Reid's principles of common sense, although like
Stewart he felt that the term was too loose and ambiguous to merit use in
philosophical discussion. McCosh saw himself to be charting a middle way
between transcendentalism/idealism on the one hand and radical empiri-
cism/utilitarianism on the other. He indicated his indebtedness to Hamilton,
but he was clearly wary of the latter's Kantianism and wished to return philosophy to what he saw as the sound and secure foundation of inductive observation of the laws of consciousness as revealed through experience of the
world—the observation of facts rather than a priori speculation.[82]

Earlier in the century, just as Princeton was turning to a conservative interpretation of Reid, the considerably more liberal Harvard Unitarians had
begun their own appropriation of Reidian thought. Upon the foundation
laid by Scottish common-sense philosophers, "Harvard professors were able
to construct a durable consensus, containing room for both Enlightenment
aspirations and Christian principles."[83] Levi Frisbie, Levi Hedge, James Walker,
and Francis Bowen, successive holders of the Alford Professorship of Natural
Religion, Moral Philosophy, and Civil Polity, all subscribed to common-sense
philosophy, forming an unbroken chain of Reidian thought at Harvard from
the turn of the century to the 1870s. Students passing through Harvard in
those years were exposed to Reid's thought in classroom lectures and in texts
either by Reid himself or his followers, including texts written by Harvard
professors.[84]

Levi Frisbie (1784–1822) became the first Alford Professor in 1817, and
his inaugural address in November of that year set the tone for the next half
century of instruction in moral philosophy at Harvard. According to Frisbie,
moral philosophy is the "science of the principles and obligation of duty." The
"unremitting labours of the moralist" are required "to relieve the sentiments
of mankind, from those associations of prejudice, of fashion, and of false
opinion, which have so constant an influence in perverting the judgment and
corrupting the heart, and to bring them back to the unbiassed dictates of
nature and common sense."[85] Levi Hedge (1766–1844) replaced Frisbie as
Alford Professor in 1827. Hedge was the author of *Elements of Logick* (1816),
a short, lucid book heavily indebted to Reid and Stewart that became an
oft-reprinted college text in the following decades. Reid and Stewart receive
citation on a number of topics throughout the book, and at the end Hedge
makes a general recommendation of the writings of Locke, Reid, Stewart,
and Brown as "compris[ing] in themselves a complete system of intellectual
philosophy."[86]

James Walker (1794–1874), a Harvard-educated pastor and editor of
American editions of Reid and Stewart, replaced Hedge in the Alford Chair
in 1839, becoming one of Harvard's most popular teachers before assuming
the presidency of Harvard in 1853. In 1834 Walker published a sermon that
exhibited affinities with Reidian thought as well as Cousinian Spiritualism.[87]
The sermon distills the potentialities of Reidian thought for religious apol-
ogetics, but there were also other factors contributing to the popularity of
common-sense philosophy among Harvard Unitarians, including its provision
of a politics and educational philosophy that consolidated the gains of the
Revolution while assuring them social stability and a dominant position in
American intellectual life.[88]

Francis Bowen (1811–1890), for his part, employed Reidian thought as a basis for maintaining a sound moral and religious perspective in the face of the rise of modern science and democratic politics. His Lowell Lectures in 1848–49 reveal a desire to harmonize religion and science on common-sense principles and to combat the infidelity and social disorder that seemed to be growing daily. Bowen endeavored to demonstrate, in his lectures, "that the fundamental doctrines of religion rest upon the same basis which supports all science, and that they cannot be denied without rejecting also the familiar truths which we adopt almost unconsciously, and upon which we depend for the conduct of life and the regulation of our ordinary concerns."[89]

One offshoot of Scottish common-sense philosophy at Harvard was the rise of New England transcendentalism.[90] A number of streams of thought fed into transcendentalism, but in many respects Reidian thought was the rock from which the movement sprang. Most of the main figures in the movement were schooled in common-sense philosophy at Harvard, and Ralph Waldo Emerson in particular seems to have been influenced by Reid and Stewart. He studied their works under Frisbie and Hedge and asserted in a prize-winning essay in 1821 that the reasonings of Reid's school "yet want the neatness and conclusiveness of a system," but "the first advance which is made must go on in the school in which Reid and Stewart have labored."[91] Emerson's thought in turn "aided Thoreau in his movement from the empirical, rationalistic version of Common Sense taught at Harvard to the espousal of idealism and the intuitive grasping of Transcendentalism," according to Richard Petersen.[92]

Petersen identifies a "moderate" school of Reidian thought in America, represented by Eliphalet Nott, president of Union College from 1804 to 1866, and Francis Wayland, who had been a student and a faculty member under Nott before becoming president of Brown in 1827. Both Nott and Wayland were exponents of common-sense philosophy in their schools, and both "were early advocates for broadening the appeal and usefulness of a college education," offering programs that were precursors of the elective system.[93] Wayland (1796–1865) wrote one of the most successful textbooks in the Reidian tradition, *Elements of Moral Science* (1835), which went through four editions in two years and by 1890 had sold 200,000 copies.[94] Although the book does not mention Reid by name, it has a clear Reidian stamp, confirmed by the fact that Wayland's *Elements of Intellectual Philosophy* (1854) was basically a gloss on Reid, citing Reid a total of 72 times.[95] *Elements of Moral Science* is divided into two parts, covering theoretical and practical ethics, and the argument is framed in analogy to the natural sciences. Wayland suggests that just as there is a preestablished physical order in the universe, so there is a preestablished moral order, and that we as humans are constitutionally equipped to make sense of that order.[96] The book presents a view of the world as a relatively unchanging entity created by God for certain ends and as such is a prime example of how common-sense philosophy was often transmuted into an ethics that, except in evangelical circles,[97] would find it difficult to withstand the onslaught of Darwinism.

Wayland's textbooks reflected the trend in American higher education, beginning around 1820, of dividing what had been called "moral philosophy" into two components—moral and mental science (or philosophy)—that reflected the faculty psychology and empirical methodology of Reid and Stewart.[98] Ezra Stiles Ely's *Conversations on the Science of the Human Mind* (1819), which recommended an inductive approach to the study of the mind on the model of Reid and Stewart, is an early example of the new "science of the mind," as is Frederick Beasley's *A Search of Truth in the Science of the Human Mind* (1822). In attempting to rehabilitate Locke, Beasley advanced one of the more searching critiques of Reid's thought to appear in early nineteenth-century America, and the book's polemical tone is indicative of the pervasive influence of Reid by the second decade of the century.[99]

The first major American textbook in mental philosophy was Francis Upham's *Elements of Intellectual Philosophy* (1827). Herbert Schneider calls it a work of "empirical psychology rather than a philosophical system," and the work has been seen by others as opening "the era of American textbooks" in psychology.[100] Upham himself saw the work to be "eclectic in character" and in fact cites a wide range of thinkers, including Buffier, Cousin, Jouffroy, and Kant. But Upham's heaviest debt is to Reid and Stewart, and he follows their ideas closely throughout the work.[101] In the decades that followed, a number of textbooks on mental philosophy were published in America. Although most of these texts were in the Reidian tradition, Schneider suggests that only McCosh "adhered closely to the Scottish school" and that as time went on, German, French, and British thought made substantial inroads on Reidian mental and moral philosophy.[102] Indeed, by the time of Noah Porter's *Human Intellect* (1868), Reid appears as but one figure in a much larger constellation of thinkers (many of them German), even though a number of Porter's ideas can be traced to Reid.[103]

But lest one assume that Reidian thought had run its course in American intellectual life, it was to be reborn once more in the pragmaticism of Charles Sanders Peirce (1839–1914), which Peirce also called "critical common-sensism." Peirce, who had been a student of Bowen at Harvard, explicitly acknowledged his debt to Reid in an era when the traditional formulations of common-sense philosophy were hardly fashionable among philosophers adjusting to the rise of Darwinism and the shattering experience of the Civil War.[104] What Peirce did, in effect, was to place some elements of Reidian thought into a Darwinian framework, making explicit its pragmatic elements and potentialities.

In published and unpublished manuscripts, Peirce dealt at length with the relationship between Reid's thought and his own. In a 1905 article published in *The Monist* he wrote, in regard to the doctrine of original beliefs maintained by "the old Scotch philosophers," that

> When I first wrote, we were hardly orientated in the new [evolutionary] ideas, and my impression was that the indubitable propositions changed with a thinking man from year to year [...] It has been only during the last two years that I have completed a provisional inquiry which shows me that the changes are so slight from generation to generation, though not imperceptible even in that short period, that I thought to own my adhesion, under inevitable modification, to the opinion of that subtle but well-balanced intellect, Thomas Reid, in the matter of Common Sense (as well as in regard to immediate perception, along with Kant).[105]

Peirce saw Reid's principles of common sense to be "the instinctive result of human experience"—vague instincts or "innate cognitive habits" adapted to a primitive environment—that serve as a foundation for higher forms of judgment and decision-making.[106] The pragmaticist's doctrine "essentially insists upon the close affinity between thinking in particular and endeavour in general. Since, therefore, action in general is largely a matter of instinct, he will be pretty sure to ask himself whether it be not the same with belief."[107] To be sure, critical common-sensism subjects instinctive beliefs to doubt, but to do so is really not that easy—the "real metal" (as opposed to "paper doubts") can only be discovered after arduously subjecting doubts to the test of practical experience.[108] Given Reid's appeal to the fruits of everyday human experience—language, common beliefs and behaviors, habits of mind—to test the claims of philosophy, it is easy to see why Peirce felt a particular affinity for Reid. In the end, Peirce finds that a critical sifting of experience "invariably leaves a certain vague residuum [of instinctive beliefs] unaffected."[109]

* * *

Reid's thought thus led in a number of directions. If it served the needs of Christianity in its confrontation with modernity, it simultaneously played an important role in the expansion of higher education and the propagation of the perspectives and methods of modern science during the nineteenth century.[110] Reidian thought promoted a vision of human nature and society that was particularly appealing to French and American educators in the aftermath of their respective political revolutions, providing a coherent and stable foundation for philosophical and scientific education in modernizing yet politically

fragile nation-states. In the German states, on the other hand, Reid's philosophical perspective clashed with the rationalistic tenor that pervaded the
mainstream of German philosophy, seeming to offer little to thinkers writing in
polities with weak middling classes, entrenched hierarchies, and lacking republican institutions and a vibrant public sphere of social and political interaction.[111] It would seem that there needs to be a "commons" for common-sense
philosophy to have much resonance, and for this reason it is not surprising that
it initially emerged and gained a foothold in Britain before spreading abroad.

By the middle of the nineteenth century, varieties of idealism, transcendentalism, positivism, and empiricism had begun to compete with Reidian
thought, but in more than a few cases these intellectual trends occurred in
tandem with, or grew from, Reidian premises. Reid himself took "removing
rubbish" and "digging for a foundation" to be one of his primary tasks,[112] and
thus it is not surprising that common-sense philosophy has proven to be fertile ground for the growth of new ideas and perspectives—if one or the other
offshoot has withered, others remain. Peirce's placement of Reid's philosophy
into an evolutionary framework indicates that Reid's thought is not exhausted
by pre-Darwinian and/or theistic understandings of the world. Moore's
influential development of common-sense philosophy similarly points to the
ongoing vitality and relevance of the thought of Reid and the common-sense
tradition, not least because, in the words of Clifford Geertz, "Common sense,
or some kindred conception, has become a central category, almost *the* central
category, in a wide range of modern philosophical systems."[113]

* * *

This relevance is nowhere more evident than in modern cognitive science,
which validates a number of Reid's ideas and general philosophical perspective. Contemporary cognitive scientists argue that the perceptual metaphors
employed by Hume and which went back to Descartes are fundamentally mistaken. As Reid recognized, our minds are *not* like a mirror, or a theater in which
images or "ideas" mingle on a stage or get projected onto a screen.[114] Second,
contemporary studies strongly suggest that there are indeed common, innate
endowments of the human mind that help us "make sense" of the world that
are not the result of experience, habit, custom, or cultural negotiation.[115] Our
mental furniture may or may not be the result of a designing Deity, but there
is nothing that restricts us from substituting the long process of evolutionary
adaptation for a creator-God working in an instant: either scenario gives us
beings well-constituted to make sense of their environments.

However we got this way, a smile has the same emotional meaning
everywhere and to everyone, as do other facial and bodily gestures, as Reid

suggested.[116] There is strong evidence from research on infants suggesting we are hardwired to perceive relationships of cause and effect, and to perceive objects in accord with the principles of cohesion, continuity, and contact: we intuitively perceive objects to move as connected, bounded units on unobstructed paths and to affect one another's motion only if they touch.[117] Hume's view of our perception of causality is thus now explicitly rejected by cognitive scientists.[118]

Noting that "the idea of cause and effect lies at the heart of both commonsense and scientific thought," researchers using sophisticated techniques of habituation and dishabituation are able to identify intuitive perceptions of causal relations of moving objects by infants. "We argue for the following perspective on the development of causality: Instead of causality being entirely the result of the gradual development of thought (Piaget, 1955; Uzgiris, 1984) or the result of prolonged experience (Hume, 1740), an important and perhaps crucial contribution is made by the operation of a fairly low level perceptual mechanism."[119] Such a mechanism operates—as Reid argued—as a basis for induction, without our knowing the inner nature of "causes" themselves:

> Perceptual input systems are required to feed central learning systems with descriptions of the environment. Such information will provide the initial data set for whatever central learning device there may be for a given domain. The input descriptions therefore must be immediately relevant to the inductive problems of that domain. If there was a "launching module," then it could provide information about the spatiotemporal and causal structure of appropriate events. And it could do this without having to know what a cause "really" is (Leslie, 1986). In short, perceptual modularity may be designed to get development started in the absence of prior relevant knowledge.[120]

Steven Pinker notes that "many cognitive scientists believe that the mind is equipped with innate intuitive theories or modules for the major ways of making sense of the world." As Pinker explains, "An intelligent system [...] cannot be stuffed with trillions of facts. It must be equipped with a smaller list of core truths and a set of rules to deduce their implications."[121] Pinker refers to these core truths and rules as features of common sense and argues that the previously dominant understanding of the mind as a "blank slate" at birth is incorrect.[122]

There is now compelling evidence that we come equipped to perceive not only causality in inanimate objects but also agency or intention in animate ones.[123] Infants distinguish between internal or animate and external causes of motion, and "this distinction emerges early: Infants are sensitive to movement patterns that correspond to animacy."[124] David Premack argues that "just as causality is the infant's principal hard-wired perception for nonself-propelled objects, so intention is its principal hard-wired perception for self-propelled objects."[125] Here too Reid's thought is suggestive; according to Reid, we are naturally constituted to perceive that we, and others, are voluntary, deliberate agents, and that human beings are the efficient causes of their actions.[126] Paul Bloom argues, in a similar vein, that babies are "natural-born dualists," that is, they naturally perceive the world to be composed of inanimate bodies and animate souls. As to the latter, "we are so hypersensitive to signs of agency that we see intention where all that really exists is artifice or accident. As Guthrie puts it the clothes have no emperor."[127]

Very young infants will react to the violation of the integrity or "identity" of discrete objects,[128] another basic aspect of an orderly and connected world. There is also mounting evidence that we are predisposed to conceive of essences;[129] Gelman and Wellman, for example, have shown that "young children distinguish insides of objects from their outsides, even when the two conflict, and believe that insides can be more essential to an object's functioning and identity."[130] Based on the evidence, they conclude that such knowledge is not learned, "because in the present studies children seem prone to believe that insides are important even when knowing little about the insides in question."[131] The essentialist mode of thought is thus a human universal, according to many developmental psychologists.[132] Reid for his part argued that although we cannot have direct knowledge of essences (just like we cannot have direct knowledge of efficient causes in nature), we do naturally perceive a difference between external attributes and internal subjects or substances of things.[133]

Echoing Reid, educational psychologists argue that "Education and development is possible because of the existence of [...] primitive 'commonsense' categories and mental operations" in children, for example, a "number sense" or the ability to judge relative magnitudes.[134] There is in fact a growing literature on children's category and concept development, which suggests that "the infants' world is not a 'blooming, buzzing, confusion.'" Rather, researchers agree that "from an early age children have the ability to form categories and use those categories to organize and make sense of the world around them."[135] These categories include "information about ontology, causation, function, intentions, and other properties that are not directly observable." Such claims build on research suggesting that "children's thought is organized into intuitive framework theories that, like scientific theories, help children organize experience, make predictions, and causally interpret their world."[136] Such theories extend to biology and go by the name of "folkbiology."[137]

Generally speaking, in contrast to strongly empiricist and behaviorist theories, cognitive scientists now argue that "[Innate] knowledge is central to common-sense reasoning throughout development. Intuitive knowledge of physical objects, people, [numerical] sets, and places develops by enrichment around a constant core, such that the knowledge guiding infant's earliest reasoning stands at the center of the knowledge guiding the intuitive reasoning of older children and adults."[138] Or as others put it, in relation to the principle of causality, "Appreciation of nonobvious causes is critical for understanding a broad range of scientific and everyday phenomena. Without the capacity to think about causes that are not directly visible, our understanding of the world would be extremely limited." In Reid's words, "Take away the light of [the] inductive principle, and Experience is as blind as a mole."[139] Contemporary researchers argue that common-sense forms of reasoning about the everyday world are inductive rather than deductive and that the "fast and frugal" heuristics of common sense are very often as effective—or even more so in some cases—in everyday problem-solving as more laborious rational processes, including multiple regression and Bayesian analysis.[140]

Just as Reid proposed the existence of a "natural language" and stressed the way in which language and linguistic structures exhibit innate features of the human mind,[141] linguists now identify an innate "universal grammar" that provides the patterns that allow human beings to construct an unconscious mental grammar for any of the languages of the world; and they suggest that all human beings come similarly equipped with a universal musical grammar, a universal visual grammar, a universal conceptual grammar, and even a universal sociocultural grammar.[142] Ray Jackendoff summarizes the data this way:

> Is there a "human nature?" the picture that has emerged is that our "human nature" consists in having a collection of innate brain specializations or modules, each of which confers on us certain kinds of cognitive powers: the ability to learn a language, to learn to appreciate music, to come to understand the visual world, to construct concrete and abstract thoughts, to learn to function in a social environment, and no doubt more.[143]

Others suggest that our moral intuitions can best be described in terms of a universal *moral* grammar.[144] The nature and scope of our moral intuitions is an area of rapidly growing interest among researchers from a variety of disciplines.[145] Probably the most important is our innate sense of reciprocity or fairness. All human beings, from a young age, make moral judgments based on principles of reciprocity, which can go either in a positive or negative direction: positive behaviors are felt to deserve a positive response, while negative behaviors deserve a negative one. This principle behind the golden rule as well as "an eye for an eye" (*lex talionis*) is clearly the foundation for more developed notions of universal justice, and very likely played a role in the evolution of altruism.[146] Economists argue that the principle of reciprocity provides an important perspective on understanding economic behavior, particularly when it comes to the ways in which people deviate from purely self-interested behavior.[147] Reid for his part spoke of a self-evident first principle of morality that sounds very much like the golden rule: "In every case, we ought to act that part towards another, which we would judge to be right in him to act toward us, if we were in his circumstances and he in ours."[148]

Other universal moral principles and values that have been identified include recognition of the binding nature of promises; human responsibility and intention; the need to redress wrongs and punish acts that threaten the group; standards of etiquette and hospitality; empathy; and generosity (particularly in leaders).[149] Ralph Linton's anthropological research suggests that universal ethical principles include the distinction between good and bad and between in-group and out-group, the regulation of sexual behavior, prohibition of incest between mothers and sons, the expectation that parents will take care of their children, the recognition of reciprocal economic obligations, and the demand for truthfulness in certain situations.[150] Does universality in itself *prove* the existence of innate mental mechanisms or principles of common sense? Of course not, but it seems highly unlikely, given the diversity of moral beliefs across cultures, that universally held moral principles would have no basis at all in our mental/emotional constitution.

Modern research thus supports Reid's suggestion that there are a number of universally self-evident principles of morality including an intuitive conception of justice.[151] No one, including Reid, suggests that these features of human nature cannot lead us astray or should rigidly determine the outcome of higher level thinking, including systems of morality.[152] Rather, an epistemology or moral system premised on the *denial* of natural and innate features of the mind is simply mistaken and potentially misleading,[153] contributing to the sense that there are no "givens," rather all is "invented" or "constructed" on the level of individual experience or sociocultural interaction. We now have pretty conclusive evidence that, as Reid long ago argued, we do in fact come equipped to "make sense" of our social and natural environments and that our most basic mental equipment is universal and cross-cultural and not determined by habit or custom. No one can credibly dispute the fact that nurture and culture play a large role in shaping and developing our mental and moral universe. The point is rather that the achievements of education and culture (including modern science) do not start from scratch but rather build upon mental and emotional foundations that have taken hundreds of thousands— even perhaps millions—of years to develop and as such are pretty stable features of our world; and that there are numerous practical implications that flow from understanding these mental foundations, including, for example, how we teach physics.[154]

* * *

Many of Reid's notions have thus aged quite well, despite the Darwinian revolution in science and the fact that scientific knowledge and research continue to become more complex, abstract, and often counterintuitive and beyond the comprehension of the average person. That is to say, despite the fact that good science operates skeptically, and that the technologies and theories of science are increasingly removed from everyday understandings and awareness, at some level there remains a link between common sense, understood as basic intuitive features of human understanding, and science, including social science.[155]

The interesting question is, what is the nature of that linkage? As we have seen, thinkers have been grappling with this question for millennia, and to this day there are a variety of positions taken on the question. During the 1950s, in the wake of the creation of the atomic bomb and the growing public awareness of the strange features of modern physics, a number of prominent thinkers addressed this topic. Writing in 1951, Jacob Bronowski presented, in *The Common Sense of Science*, a somewhat unsettling picture of the relationship between the two entities. His fundamental argument is that quantum mechanics has permanently upended the classical notion of causality as a logical, mechanical chain of causes and effects—"so natural and obvious to us"—as a goal and outcome of scientific inquiry. For Bronowski, the uncertainty principle has taken the metaphysics but not the order out of science; science is not about causes but about describing the world "in an orderly scheme or language which will help us to look ahead […] This is a very limited purpose. It has nothing whatever to do with cause and effect at all, or with any other special mechanism." Indeed, all general metaphysical beliefs, like the belief that "the rules by which we are acting are universal" are "at odds with the principles of science."

After making these astonishing claims, Bronowski then heads into the safer territory that Robert Oppenheimer would go on to explore in *Science and the Common Understanding*, first published in 1953: "causal laws" at higher levels are in fact accumulations of the "laws of chance" at the atomic level. In emphasizing the disjuncture between the understandings of classical and modern physics, however, Bronowski ignores early-modern scientific nominalism and skepticism about our ability to "know" causes beyond the laws or rules by which they operate; and he gives scant attention to the question of how our intuitive expectation and understanding of causality is connected to scientific endeavor, if science itself has nothing to do with causality.[156]

Oppenheimer, for his part, advanced the idea that science and common sense are complementary rather than contradictory modes of thinking, occupying different levels of understanding. While scientific knowledge of the behavior of atomic particles can seem to violate our everyday understandings of how the world works, it is relevant only on scales that transcend the frame of everyday experience—for example, the very large and the very small. And our scientific understanding of the atomic realm is known to us by "reducing the experience with atomic systems to experiment and observation made manifest, unambiguous, and objective in the behavior of large objects, where the precautions and incertitudes of the atomic domain no longer apply." Thus, "the measurements that we have talked about in such highly abstract form do in fact come down in the end to looking at the position of a pointer, or the reading of time on a watch, or measuring out where on a photographic plate or a phosphorescent screen a flash of light or a patch of darkness occurs." Consequently, common sense is not wrong when it comes to understanding the world of everyday experience, and indeed science takes many of its conclusions for granted; "common sense is wrong only if it insists that what is familiar must reappear in what is unfamiliar." Oppenheimer thus provides a nuanced perspective on the "deep, intimate, and subtle" relations between the findings of modern science and human beings' everyday intuitions and perceptions of the physical world.[157]

In *Common Sense and Science*, published in the same year as Bronowski's book (1951), James Conant looked to fields other than physics, where the links between common sense and science are more clear. Conant asserted that much of the "conceptual scheme" of science is the same as that of common sense, for example, that objects exist in three-dimensional space independent of the observer, that nature acts in a uniform and regular manner, that there are other personalities with whom we can communicate, and so on. As such, "common sense notions provide the stable platform on which we build." Experimental science can be thought of as "an activity which increases the adequacy of the concepts and conceptual schemes which are related to certain types of perception and which lead to certain types of activities; it is one extension of common sense." Thus, "there is a continuous gradation between the simplest rational acts of an individual and the most refined scientific experiment." To be sure, Conant recognizes that many scientific theories end up being quite "distant" from common sense. Nevertheless, he emphasizes the degree to which scientific experimentalism draws upon common-sense intuitions about the world and also the ways in which scientific method parallels everyday problem-solving.[158]

Another work from the era that dealt with these issues is *The Foundations of Common Sense* ([1949] 1999) by Nathan Isaacs. Similar to Oppenheimer and Conant, Isaacs argues that "The scientific view of the world, in its different aspects, has *developed continuously* out of the naïve perceptual one, still leans on this, and retains a number of fundamental features in common with it." According to Isaacs,

> The unbreakable link between our naïve common sense world-picture and our subsequent scientific one is the special dominant status of our perceptual experience [...] whatever science, i.e. scientific method, may ultimately tell us about the nature of our physical world, this [cannot] but be compatible with the acceptance of the controlling status of perception as the key to that world, since the ultimate test is always agreement of the theory with the facts of observation and experiment (or rather, the observed outcome of experiment). Primary perceptual experience thus fully retains its place as final arbiter, as it were for and on behalf of the objective world.[159]

A more recent work that picks up and develops these themes in great detail is John Ziman's *Real Science: What It Is, and What It Means* (2000). Ziman puts the development of human cognition as it relates to science in an evolutionary framework, arguing that our minds are uniquely adapted to make sense of the world. "Our elementary representations of the life-world must correspond well enough with that world to have been naturally selected and re-selected in innumerable close encounters with it. In effect, the entities about which we have knowledge have been actively involved in the making of that knowledge." Thus, our basic "life-world knowledge" brings with it an "exceptional epistemic authority—an *authenticity* that puts it beyond all reasonable doubt." While this knowledge is selected for prediction and control of our environment and not scientific understanding, common sense and science are intimately related and in fact "define each other, as they evolve together." Scientific knowledge cannot be formally constituted "as a separate, self-contained epistemic domain. Even its grandest theoretical paradigms are inferred from and rooted in down-to-earth empirical 'facts' and always have to incorporate uncritically a great deal of 'taken for granted' life-world knowledge."[160] Science is thus not simply a systematic enlargement of "folk science." On the contrary, "it is in its conformity to the *small* print of common sense that science is distinctive." That is, the cognitive norms of science such as "accuracy, specificity, reproducibility, generality, coherence, consistency, rigour, and so on—are all perfectly commonsensical: but they are seldom applied simultaneously outside science." In arriving at theories that may contradict mundane beliefs or what Reid called "common understanding," scientific reasoning "operates as an adjunct to ordinary modes of practical reasoning in revising or extending 'common sense.'"[161]

Ziman thus presents a view in which science is firmly rooted in the basic features of common sense, making rational use of them to produce what can initially seem to be counterintuitive knowledge but which over time comes to be reassimilated to everyday understanding and the intuitions and perceptions out of which it grew. Except for Ziman's incorporation of evolutionary psychology, this perspective aligns with Reid's view, concisely expressed by Reid in the metaphor of a tree: "All that we know of nature, or of existences, may be compared to a tree, which hath its root, trunk, and branches. In this tree of knowledge, perception is the root, common understanding is the trunk, and the sciences are the branches."[162]

* * *

There is thus a long tradition of thinkers who suggest that there are important and ineluctable links between common sense and the theory and practice of science; and this helps to explain why scientists often make rhetorical use of "common sense," as when T. H. Huxley stated that "science is nothing but trained and organized common sense,"[163] or why Charles Darwin, who in this theory and use of language was a realist, used "plain English" and everyday images to conceive and explain his theory and persuaded his peers by appeals to common sense, urging "that we can trust a theory which explains so many large classes of facts because this 'is a method used in judging in the common events of life.'"[164] Contemporary popular science writers invoke "common sense" when countering the claims of pseudo-science,[165] while the authors of the *Dictionary of Science for Everyone* suggest that "Newton's three laws [of motion] seem quite simple and commonsensical—which they are" and demonstrate at length the links between complex, abstract theories and everyday understandings.[166]

Even in the realm of pure mathematics, practitioners argue that it is incorrect to think that mathematics is a closed, purely logic-driven enterprise; Lord Kelvin famously stated that it is wrong to imagine that mathematics is "repulsive to common sense. It is merely the etherialization of common sense."[167] Davis and Hersh argue that "Even for a very tiny piece of math, the task of giving an absolutely air-tight formal proof turns out to be amazingly complicated. Professedly vigorous proofs usually have holes that are covered over by intuition." Mathematical argument is a form of social interaction in which "'proof' is a complex of the formal and informal, of calculations and casual comments, of convincing arguments and appeals to the imagination and the intuition." The point of a proof is to convince the intended audience, "a group of professionals with training and mode of thought comparable to that of the author. Consequently, our confidence in the correctness of our results is not absolute, nor is it fundamentally different in kind from our confidence in our judgements of the realities of ordinary life."[168] Reid said much the same thing about mathematics, characteristically balancing individual, intuitive judgment with those of a larger community:

> Society in judgement, of those who are esteemed fair and competent judges has effects very similar to those of civil society; it gives strength and courage to every individual; it removes that timidity which is as naturally the companion of solitary judgement, as of a solitary man in the state of nature. Let us judge for ourselves, therefore, but let us not disdain to take that aid from the authority of other competent judges, which a mathematician thinks it necessary to take in that science, which of all sciences has least to do with authority.[169]

There is thus in principle a continuum stretching between common sense and science, because as Reid and the mathematicians just cited suggest, determining what counts as proper intuitive judgment is to some degree dependent on intersubjective confirmation—a point that Ziman argues at length.[170] It seems clear then that common sense and science are linked in the smaller publics of practicing scientists and in the larger public sphere of popular science, and in the end it may be impossible to separate the two. Psychologists, certainly, are now arguing that "commonsense concepts cannot easily be transcended or substituted for by theoretical terms." Theoretical, "scientific" terms are them-selves dependent upon commonsense terms for definition: "Commonsense is an extremely complex structure playing a vital role in every step of theory construction and empirical validation in psychology."[171] Thomas Luckmann sums up this general line of thinking very well:

> Science, by its empiricist principles of verification (or corroboration) is knowledge that pertains exclusively to a reality whose manifestations are accessible to everyone in the ordinary world of everyday life, no matter how far removed its concepts are from the level of ordinary experience and how incomprehensible its formulae are to common sense. The "home base" of modern science is also the "home base" of common sense.[172]

The Humean position, it may be said, roughly parallels the position of "postmodern" thinkers in the humanities and social sciences who base their thinking on a far-reaching transcendental critique of common-sense notions about our ability to know and represent the "real," whether it be a coherent self or an objective external reality.[173] Hume was by all accounts a man of letters rather than a natural philosopher, and despite his Newtonian rhetoric his thinking took place primarily in the world of classical learning, politics, the arts, and letters.[174] Reid, on the other hand, was nurtured in, and imbibed deeply, Newtonian natural philosophy—he was a practicing natural philosopher and mathematician; he rubbed shoulders with a wide range of physicians, mathematicians, and natural scientists; and his thinking took place in a Baconian/Newtonian experimental matrix (see Chapter 8). Just like present-day postmodernists, Hume felt that our intuitive, common-sense judgments are fundamentally misleading, constructed rather than given, and progress is made only by taking a deconstructive theoretical leap beyond them. And like many present-day scientists, Reid felt that intuitive judgments are the roots of useful and as far as we are concerned "objective" knowledge. His own philosophical analysis revealed both the limits of that kind of analysis and that the world is out there—and "given" to us—to be investigated, and if we do it right we will find out things about it that can be confirmed by anyone else, anywhere else in the world, and it's just silly and patently absurd, an abuse of language, to speak otherwise.

It can thus be argued that the Hume–Reid debate is a productive one as long as each side acknowledges that what is being said emerges out of quite different sets of practices with very different agendas. Skepticism and foundationalism need each other if neither of them is to go off the rails. If on the one hand the natural sciences have often proceeded as if scientists were merely reading a script written in the hand of nature, on the other hand thinkers in the humanities and social sciences can just as easily fall prey to fanciful notions that hover uncertainly between an exalted sense that we make everything up ourselves and contrition over our inability to be sure about anything at all. Both extremes tend to be based on a denial of the common-sense roots of our knowledge. The former forgets that advanced concepts emerge out of a matrix of individual *and* social experience—as Thomas Kuhn famously argued[175]—and hence that any bit of "knowledge" we gain is textured by the psychosocial process in which we gain it. Too much faith is put in our perceptual equipment and the conclusions it can produce, without paying attention to the *process*, including the role other human beings play in shaping and confirming our expectations. The "common" in common sense gets forgotten.

The latter for their part lose faith entirely in the "sense" part of the entity, laying too much stress on the process and not enough on the very reliable knowledge provided by the elemental building blocks of that process—the "givens" which allow us to get started at all. Skepticism about the absolute objectivity of our knowledge is certainly justified—Reid himself felt that we cannot "know," in an absolute sense, the nature of efficient causes, but so is the recognition that such skepticism is productive only when it has something to work against; and that "something" is none other than our commonsensical intuitions which, in their turn, are mediated, corrected, and justified within the larger human community.

NOTES

Introduction

1 To that end, the Epilogue carries the story forward into the present, noting both the development of Reidian ideas in Europe and America and how they have fared in light of the findings and theories of modern physics, philosophy of science, and cognitive science.

2 Nicholas Rescher, *Common Sense: A New Look at an Old Philosophical Tradition* (Marquette: Marquette University Press, 2005), 12.

3 Lewis Wolpert, *The Unnatural Nature of Science* (Cambridge, MA: Harvard University Press, 1993), xi–xii.

4 Ibid., 106.

5 Alan Cromer, *Uncommon Sense: The Heretical Nature of Science* (New York: Oxford University Press, 1993), 3.

6 This point is discussed and documented at length in the Epilogue. For a recent overview of the scientific research on this topic, see Paul Bloom, *Descartes' Baby: How the Science of Child Development Explains What Makes Us Human* (New York: Basic Books, 2004).

7 See, for example, Nassim Nicholas Taleb, *Fooled by Randomness: The Hidden Role of Chance in Life and in the Markets* (New York: Random House, [2001] 2005); *The Black Swan: The Impact of the Highly Improbable* (New York: Random House, [2007] 2010).

8 Rescher, *Common Sense*, 31.

9 Ibid., 158, 165.

10 Ibid., 206–8, 235. Similar arguments are presented in Mark Kingwell, "The Plain Truth about Common Sense: Skepticism, Metaphysics, and Irony," *Journal of Speculative Philosophy*, 9, no. 3 (1995): 169–88.

11 John Ziman, *Real Science: What It Is, and What It Means* (Cambridge: Cambridge University Press, 2000), 314.

12 Ibid., 313. A similar perspective is advanced in Nathan Isaacs, *The Foundations of Common Sense: A Psychological Preface to the Problems of Knowledge* (London: Routledge, [1949] 1999), 52–59. Isaacs argues that our primary, commonsensical "perceptual experience" is both the source and test of our higher-order scientific knowledge. Discussed further in the Epilogue.

13 Michael Dummett, "Common Sense and Physics," in *Perception and Identity: Essays Presented to A.J. Ayer, and His Replies*, ed. G. F. Macdonald (Ithaca, NY: Cornell University Press, 1979), 24.

14 Nick Huggett, "Identity, Quantum Mechanics, and Common Sense," *The Monist*, 80, no. 1 (1997): 118.

15 Ibid., 127–28.

16 Peter Forrest, "Common Sense and a 'Wigner-Dirac' Approach to Quantum Mechanics," *The Monist*, 80, no. 1 (1997): 131–59.

17 Nicholas Maxwell, "Physics and Common Sense," *British Journal for the Philosophy of Science*, 16, no. 64 (1966): 295–311.

18 Bertrand Russell, *The ABC of Relativity* (Abingdon and New York: Routledge, 2009).

19 James B. Conant, *Science and Common Sense* (New Haven, CT: Yale University Press, 1951); J. Robert Oppenheimer, *Science and the Common Understanding* (New York: Simon and Schuster, 1954).

20 Kathleen Wilkes, "The Relationship between Scientific Psychology and Common Sense Psychology," in *Folk Psychology and the Philosophy of Mind*, ed. Scott Christensen and Dale Turner (Hillsdale: Lawrence Erlbaum Associates, 1993), 167–87.

21 Rescher, *Common Sense*, 144–45.

22 Daniel Kahneman, *Thinking, Fast and Slow* (New York: Farrar, Straus, and Giroux, 2011), 13.

23 Ibid., 25.

24 Ibid., 31.

25 Ibid., 21, 25.

Chapter 1 Common Sense and Scientific Thinking before Copernicus

1   See, for example, *A Short History of Philosophy*, by Robert C. Solomon and Kathleen Higgins (New York: Oxford University Press, 1996). This historical survey of philosophy makes the opposition between philosophical and everyday knowledge one of the guideposts of its narrative.

2   Ibid., 27–38; quote from 40.

3   As he puts it in the *Phaedo*, "above all other men," true philosophers scorn the indulging of the body. *The Dialogues of Plato Vol. 1*, trans. B. Jowett (New York: Random House, 1937), 448.

4   Ibid., 448–49, 783–84.

5   Ibid., 412–13.

6   Ibid., 739.

7   Ibid., 740–43.

8   Ibid., 752.

9   Ibid., 753.

10 Ibid., 754.

11 Ibid., 757. In the *Phaedrus*, Socrates speaks of an upper world of absolute truths that only philosophers have the wings to attain, and a lower world inhabited by nonphilosophers that is characterized by "confusion and perspiration and extremity of effort" whose occupants "feed upon opinion." Ibid., 252.

12 This is Jowett's formulation. See *The Dialogues of Plato Vol 2*, trans. B. Jowett (New York: Random House, 1937), 848. As Socrates puts it in *Republic*, "Dialectic, and dialectic alone, goes directly to the first principle and is the only science which does away with hypotheses in order to make her ground secure; the eye of the soul, which is literally buried in an outlandish slough, is by her gentle aid lifted upwards." *The Dialogues of Plato Vol 1*, trans. B. Jowett (New York: Random House, 1937), 793.

13 Ibid., 794.

14 Ibid., 360–61.

15 Ibid., 253.

16 Ibid., 365–66.

17 See, for example, Arlene W. Saxonhouse, "The Socratic Narrative: A Democratic Reading of Plato's Dialogues," *Political Theory*, 37, no. 6 (2009): 728–53; Christina Tarnopolsky, "Platonic Reflections on the Aesthetic Dimensions of Deliberative Democracy," *Political Theory*, 35, no. 3 (2007): 288–312.

18 As noted in the Introduction, I will use a variety of related terms in this study to specify more clearly the various meanings of "common sense." "Common-sense *experience*" seems to capture the strong experiential element of Aristotle's epistemology, but I will also use the term whenever I want to emphasize the experiential elements of common sense. I will also use the terms "common sense," "common experience," "common knowledge," "everyday experience," "common opinion," and "intuition," among other terms, as required by the context. "Common sense" will ordinarily be used to refer to the commonly held, seemingly self-evident perceptions and judgments of the average person, and hence of the local and/or larger human community. Since the term carries with it both communal ("common") and perceptual ("sense") connotations, as well as a strong implication of intuitive or self-evident knowledge, its meaning will necessarily be somewhat elastic, emerging out of the particular context in which it is used.

19 Helpful discussions of Aristotelian thought from this perspective include Peter Dear, *Discipline and Experience: The Mathematical Way in the Scientific Revolution* (Chicago: University of Chicago Press, 1995), 3–30; Lorraine Daston, "Baconian Facts, Academic Civility, and the Prehistory of Objectivity," *Annals of Scholarship*, 3, no. 4 (1991): 337–64; Scott Atran, *Cognitive Foundations of Natural History: Towards an Anthropology of Science* (Cambridge: Cambridge University Press, 1990), Part II. *The Cambridge Companion to Aristotle* notes that Aristotle's science "is commonsensical. It tries to explain the general structure and functioning of the world in terms of processes whose operations are evident to all; it involves no theoretical arcana." R. J. Hankinson, "Science," in *The Cambridge Companion to Aristotle*, ed. Jonathan Barnes (Cambridge: Cambridge University Press, 1995), 144.

20 *The Basic Works of Aristotle*, ed. Richard McKeon (New York: Random House, 1941), 9–10.

21 Atran, *Cognitive Foundations of Natural History*, 88.

22 Aristotle especially emphasizes this point in the *Posterior Analytics*. See McKeon, *The Basic Works of Aristotle*, 110–24.

23 Ibid., 943–44. See also 937–38.

24 Ibid., 237.

25 Ibid., 110.

26 Atran, *Cognitive Foundations of Natural History*, 92–93.

27 McKeon, *The Basic Works of Aristotle*, 184–86. See also Robin Smith, "Logic," in *The Cambridge Companion to Aristotle*, ed. Jonathan Barnes (Cambridge: Cambridge University Press, 1995), 49–51.

28 McKeon, *The Basic Works of Aristotle*, 937–38.

29 See Dear, *Discipline and Experience*, 3–30; Daston, "Baconian Facts," 337–64; J. Milton, "Induction before Hume," *British Journal for the Philosophy of Science*, 38 (1987): 49–74. As F. J. Hankinson puts it, "For Aristotle explanation is general. He is not really concerned with the particular causes of individual events, but with the general patterns which run invariably (or at least for the most part) through the structure of the world." *The Cambridge Companion to Aristotle*, ed. Jonathan Barnes (Cambridge: Cambridge University Press, 1995), 117.

30 Ilpo Halonen, Jaako Hintikka, "Aristotelian Explanations," *Studies in the History and Philosophy of Science*, 31, no. 1 (2000), esp. 125–28; Robin Smith, "Logic," in *The Cambridge Companion to Aristotle*, ed. Jonathan Barnes (Cambridge: Cambridge University Press, 1995), 47; R. J. Hankinson, "Philosophy of Science," *The Cambridge Companion to Aristotle*, ed. Jonathan Barnes (Cambridge: Cambridge University Press, 1995), 110; Max Hocutt, "Aristotle's Four Becauses," *Philosophy*, 49/190 (1974): 385–99; R. L. Woolhouse, *The Empiricists* (Oxford: Oxford University Press, 1988), 6–7; Richard Campbell, *Truth and Historicity* (Oxford: Clarendon Press, 1992), 147–48.

31 McKeon, *The Basic Works of Aristotle*, 240–41 (*Physics*); 690–93, 712 (*Metaphysics*).

32 Ibid., 1024–25.

33 Atran, *Cognitive Foundations of Natural History*, 85.

34 Ibid., 98–99.

35 Ibid., Part I.

36 Ibid., 121–22.

37 McKeon, *The Basic Works of Aristotle*, 432–35.

38 Ibid., 353–54.

39 H. Carteron, quoted in Hankinson, "Science," 146.

40 Isaac Newton, "De Motu Corporum," in D. T. Whiteside, ed., *The Mathematical Papers of Isaac Newton Vol. VI, 1684–91* (Cambridge: Cambridge University Press, 1974), 95.

41 Hankinson, "Science," 144–50.

42 See, for example, *Posterior Analytics*, in McKeon, ed., *The Basic Works of Aristotle*, 120–24.

43 See Charles H. Kahn, "Sensation and Consciousness in Aristotle's Psychology," *Articles on Aristotle Vol. 4*, ed. Jonathan Barnes (London: Duckworth, 1978), 1–31. Similarly, Pavel Gregoric argues that "If we decide that the monitoring function of the common sense is a kind of consciousness, it is a consciousness of perceptual states only." Pavel Gregoric, *Aristotle on the Common Sense*. (Oxford: Oxford University Press, 2007), 211. My interpretation of Aristotle's concept of common sense is primarily based on Kahn; on my own reading of Aristotle's *On the Soul*, Books II and III; and on Steven Everson, "Psychology," in *The Cambridge Companion to Aristotle*, ed. Jonathan Barnes (Cambridge: Cambridge University Press, 1995), esp. 187–88.

44 Quoted in Charles H. Kahn, "Sensation and Consciousness in Aristotle's Psychology," 13.

45 Gregoric, *Aristotle on the Common Sense*, 206.

46 Ibid., 12.

47 Such writers included Herophilus, Erasistratus, Posidonius, Galen, Augustine, and Nemesius. See Walther Sudhoff, *Die Lehre von den Hirnventrikeln in textlicher und graphischer Tradition des Altertums und Mittelalters* (Leipzig: Verlag von Johann Ambrosius Barth, 1913); Harry Austin Wolfson, "The Internal Senses in Latin, Arabic, and Hebrew Philosophic Texts," *Harvard Theological Review*, 27, no. 2 (April 1935): 69–133; Simon Kemp, *Medieval Psychology* (New York: Greenwood Press, 1990), esp. Chapter 4; A. Mark Smith, "Getting the Big Picture in Perspectivist Optics," in *The Scientific Enterprise in Antiquity and the Middle Ages* (Chicago: University of Chicago Press, 2000), 315–36.

48 Kemp, *Medieval Psychology*, 36. Augustine, Avicenna, Averroes, al-Ghazali, Albertus Magnus, Thomas Aquinas, Wilhelm von Saliceto, Lanfranc, Heinrich von Mondeville, Roger Bacon, and Guy de Chauliac, among many others, place the *sensus communis* in the front ventricle of the brain. See Sudhoff, *Die Lehre von den Hirnventrikeln*.

49 Smith, "Getting the Big Picture in Perspectivist Optics," 319.

50 Kemp, *Medieval Psychology*, 55. Smith concurs that the Aristotelian legacy is "clear" in this model, but he also discusses its departures from Aristotle in some detail. See Smith, "Getting the Big Picture in Perspectivist Optics," 320–21.

51 See Thomas Aquinas, *Summa Theologica*, Part I, VI. Man, Questions 78, 79, 84, and 87, in Anton C. Pegis, ed., *Basic Writings of Saint Thomas Aquinas Volume One* (New York: Random House, 1945).

52 Edward Grant, *The Foundations of Modern Science in the Middle Ages: Their Religious, Institutional, and Intellectual Contexts* (Cambridge: Cambridge University Press, 1996); Charles B. Schmitt, *Aristotle and the Renaissance* (Cambridge: Harvard University Press, 1983).

53 Grant, *The Foundations of Modern Science in the Middle Ages*, 161.

54 Quoted in Grant, *The Foundations of Modern Science in the Middle Ages*, 69.

55 Ibid., 69.

56 Grant states: "Aristotle's natural books formed the basis of natural philosophy in the universities, and the way in which medieval scholars understood the structure and operation of the cosmos must be sought in those books. By his use of assumptions, demonstrated principles, and seemingly self-evident principles, Aristotle imposed a strong sense of order and coherence on an otherwise bewildering world" (ibid., 54).

57 See, for example, Michael H. Shank, *The Scientific Enterprise in Antiquity and the Middle Ages* (Chicago: University of Chicago Press, 2000).

58 Leonardo da Vinci, for example, wrote extensively about *senso comune* in the context of his theory of painting and perspective. "He describes the light-filled process of the real vision from which painting arises as follows: 'the eye receives the species, or similitudes of objects and [the similitudes pass] from there to the *impressiva*, and from this *impressiva* to the common sense (*senso comune*), and there it is judged.'" David Summers, *The Judgment of Sense: Renaissance Naturalism and the Rise of Aesthetics* (Cambridge: Cambridge University Press, 1987), 71.

Chapter 2 The Challenge of Modern Science and Philosophy

1 See Steven Shapin, *The Scientific Revolution* (Chicago: University of Chicago Press, 1996), 25–28, for discussion of the challenges posed by Copernicanism to common-sense experience. For the clash between Copernicanism and biblical cosmology, see Kenneth J. Howell, *God's Two Books: Copernican Cosmology and Biblical Interpretation in Early Modern Science* (Notre Dame: University of Notre Dame Press, 2002).

2 See *Calvin: Commentaries*, trans. and ed. Joseph Haroutunian (Philadelphia: Westminster Press, 1958), 29–35, 356–57; and "Letter to the Grand Duchess Christina," in *Discoveries and Opinions of Galileo*, trans. and ed. Stillman Drake (New York: Anchor Books, 1957), 175–216, esp. 181–82.

3 Howell, *God's Two Books*.

4 Nicolaus Copernicus, *On the Revolutions of the Heavenly Spheres* (Amherst: Prometheus Books, 1995), 5.

5 Quoted in Barbara Beinkowska, "The Reception of the Heliocentric Theory in Polish Schools in the 17th and 18th Centuries," in *The Reception of Copernicus' Heliocentric Theory*, ed. Jerzy Dobryzycki (Dordrecht: D. Reidel, 1973), 105.

6 See, for example, Thomas Kuhn, *The Copernican Revolution: Planetary Astronomy in the Development of Western Thought* (Cambridge: Harvard University Press 1957, 1985), Chapter 6, "The Assimilation of Copernican Astronomy." For detailed treatment

of this topic see Jerzy Dobryzycki, ed., *The Reception of Copernicus' Heliocentric Theory* (Dordrecht: D. Reidel, 1973).

7   Quoted in Kuhn, *The Copernican Revolution*, 190.

8   Ibid., 189.

9   Copernicus, *On the Revolutions of the Heavenly Spheres*, 17.

10   See Galileo Galilei, *Dialogue Concerning the Two Chief World Systems—Ptolemaic and Copernican*, trans. Stillman Drake (Berkeley: University of California Press, 1967), Second Day and passim.

11   Ibid., 41.

12   Simplicius repeatedly makes the point that the Copernican theory is a violation of common-sense experience and hence represents a departure from the science of Aristotle. Much of the *Dialogue* consists in arguments responding to this fundamental charge. See, for example, ibid., 32, 34, 46, 248, 254, 256.

13   Montaigne, Charron, and Camus, among other skeptics, cited Copernicanism as helping to cast doubt on our ability to attain certain knowledge. See Richard H. Popkin, *The History of Scepticism from Erasmus to Descartes* (Assen: Van Gorcum, 1960), 51, 61, and 98–99. Popkin concludes that

> The "new science" of Copernicus, Kepler, Galileo and Gassendi had "cast all in doubt." The discoveries of the New World and in the classical world had given other grounds for scepticism. And the "nouveaux pyrrhoniens" showed man's inability to justify the science of Aristotle, of the Renaissance naturalists, of the moralists, and of the scientists as well [...] The *crise pyrrhonienne* had overwhelmed man's quest for certainty in both religious and scientific knowledge. (Ibid., 111)

Galileo's contemporary John Donne asserted in his famous poem "Anatomy of the World" (1611) that "The new Philosophy calls all in doubt, / The Element of fire is quite put out; / The Sun is lost, and th' earth, and no man's wit / Can well direct him where to look for it... / 'Tis all in pieces, all coherence gone." Quoted in Kuhn, *The Copernican Revolution*, 194.

14   Johannes Kepler, *Epitome of Copernican Astronomy & Harmonies of the World*, trans. Charles Glenn Wallis (Amherst, NY: Prometheus Books, 1995), 14.

15   John Wilkins, *The Discovery of a New World; or, A Discourse Tending to Prove, That (It Is Probable) There May Be Another Habitable World in the Moon* (1638), in *The Mathematical and Philosophical Works of the Right Rev. John Wilkins* (London: Frank Cass, 1970), 11.

16   John Wilkins, *A Discourse Concerning a New Planet, Tending to Prove, That (It Is Probable) Our Earth Is One of the Planets*, in Wilkins, *The Mathematical and Philosophical Works*, 137. Going further, Wilkins states that since "It is plain that common people judge by their senses [...] their voices are altogether unfit to decide any philosophical doubt, which cannot well be examined or explained without discourse and reason" (138). Shapin presents other examples of this sort of elitist rhetoric from the era. See Shapin, *The Scientific Revolution*, 94, 122–23.

17   Wilkins, *A Discourse Concerning a New Planet*, 209. Wilkins goes on to discuss the ship analogy as a way of understanding the possibility that the earth may be in motion without our knowing it (210–11).

18   René Descartes, "Rules for the Direction of the Mind," in *The Philosophical Writings of Descartes Vol. 1*, trans. John Cottingham, Robert Stoothoff, Dugald Murdoch

(Cambridge: Cambridge University Press, 1985), 41–42; William Harvey, *Anatomical Exercitations, Concerning the Generation of Living Creatures* (London, 1653), Exercise 57, pp. 350–51; Kepler, *Epitome of Copernican Astronomy*, 15.

19 Thomas Hobbes mocked the Scholastic model of the mind thus: "Some say the Senses receive the Species of things, and deliver them to the Common-sense; and the Common Sense delivers them over to the Fancy, and the Fancy to the Memory, and the Memory to the Judgement, like handing of things from one to another, with many words making nothing understood." Thomas Hobbes, *Leviathan* (New York: Penguin Classics, 1985), 93.

20 For a helpful discussion of Locke's theory of mind, see Vere Chappell, "Locke's Theory of Ideas," in *The Cambridge Companion to Locke*, ed. Vere Chappell (Cambridge: Cambridge University Press, 1994), 26–55. See also Roger Woolhouse, "Locke's Theory of Knowledge," in *The Cambridge Companion to Locke*, 146–71.

21 See Peter Dear, *Discipline and Experience: The Mathematical Way in the Scientific Revolution* (Chicago: University of Chicago Press, 1995).

22 Quotes are from John Wilkins, *A Discourse Concerning a New Planet*, 138, and "To the Reader" (n.p.).

23 See the introductory material and Book I of Francis Bacon, *The New Organon*, ed. Lisa Jardine (Cambridge: Cambridge University Press, 2000).

24 Ibid., 41.

25 Ibid., 17.

26 Ibid., 37–38.

27 Ibid., 17.

28 Ibid., 48.

29 Ibid., 29–30.

30 Ibid., 58; see also Bacon's well-known comments in *The New Atlantis* about science as the search for causes, in *Francis Bacon: A Critical Edition of the Major Works*, ed. B. Vickers (Oxford: Oxford University Press, 1996), 480. See also R. S. Woolhouse, *The Empiricists* (Oxford: Oxford University Press, 1988), 19; and Margaret Osler, "John Locke and the Changing Ideal of Scientific Knowledge," in *Philosophy, Religion, and Science in the Seventeenth and Eighteenth Centuries*, ed. John Yolton (Rochester: University of Rochester Press, 1990), 4–5.

31 Lorraine Daston, "Baconian Facts, Academic Civility, and the Prehistory of Objectivity," *Annals of Scholarship*, 3, no. 4 (1991): 345.

32 Shapin, *The Scientific Revolution*, 101–17. See also Kenneth Clatterbaugh, *The Causation Debate in Modern Philosophy 1637–1739* (New York: Routledge, 1999), Chapter 7; and Margaret Osler, "John Locke and the Changing Ideal of Scientific Knowledge," in *Philosophy, Religion, and Science in the Seventeenth and Eighteenth Centuries*, 325–38.

33 Robert Boyle, "Of the Excellency and Grounds of the Corpuscular or Mechanical Philosophy," in *The Works of Robert Boyle Vol. IV*, ed. T. Birch (Hildesheim: G. Olms Verlagsbuchhandlung, 1966), 69.

34 For example, Pierre Gassendi, in his *Exercitationes Paradoxicae adversus Aristoteleos* of 1624, argued that our knowledge of the world is restricted to outward appearances of things, rather than inner essences and infallible causes, and hence "science" in the Aristotelian sense is impossible. See Popkin, *The History of Scepticism from Erasmus to Spinoza*, 100–102.

35 Dear, *Discipline and Experience*; Steven Shapin, *A Social History of Truth: Civility and Science in Seventeenth-Century England* (Chicago: University of Chicago Press, 1994); Mary

Poovey, *A History of the Modern Fact: Problems of Knowledge in the Sciences of Wealth and society* (Chicago: University of Chicago Press, 1998), 191.

36   Dear, *Discipline and Experience*; Daston, "Baconian Facts," 343.

37   According to Stillman Drake, Galileo felt

> that in order to become science, philosophy must throw out blind respect for authority, but he also saw that neither observation, nor reasoning, nor the use of mathematics could be thrown out along with this. True philosophy had to be built upon the interplay of all three [...] He knew very well that the unsupported evidence of the senses might lead a man astray.

Drake suggests further that Galileo "realized that philosophy must learn to be content with pursuing limited objectives, reaching out gradually into the infinity of unknown events and undiscovered laws of nature, without ever achieving complete and exact knowledge of anything at all." *Discoveries and Opinions of Galileo*, 223–24. Peter Machamer has argued that "Galileo used a comparative, relativized geometry of ratios as the language of proof and mechanics, which was the language in which the book of nature was written [...] Using this geometry one does not look for physical constants or solutions to problems in terms of absolute numerical values." Peter Machamer, "Galileo's Machines, His Mathematics, and His Experiments," in *The Cambridge Companion to Galileo*, ed. Peter Machamer (Cambridge: Cambridge University Press, 1998), 65. In the Second Day of the *Dialogue* Galileo has Salviati say that

> We do not really understand what principle or what force it is that moves stones downward, any more than we understand what moves them upward after they leave the thrower's hand [...] We have merely [...] assigned to the first the more specific and definite name 'gravity,' whereas to the second term we assign the more general term 'impressed force. (*Dialogue Concerning the Two Chief World Systems*, 234)

38   See Douglas Jesseph, "Hobbes and the Method of Natural Science," in *The Cambridge Companion to Hobbes*, ed. Tom Sorrell (Cambridge: Cambridge University Press, 1996), 86–91.

39   Thomas Hobbes, *Leviathan* (New York: Penguin Classics, 1985), 183.

40   See Dear, *Discipline and Experience*, Chapter 8 and Conclusion.

41   For more, see Osler, "John Locke and the Changing Ideal of Scientific Knowledge;" Stephen Buckle, "British Sceptical Realism: A Fresh Look at the British Tradition," *European Journal of Philosophy*, 7, no. 1 (1999): 1–29; Clatterbaugh, *The Causation Debate in Modern Philosophy*, 186–91.

42   John Locke, *An Essay Concerning Human Understanding*, 6th ed. (Amherst: Prometheus Books, 1995), dedication, p. xi.

43   Ibid., I.i.4, p. 2.

44   Ibid., I.ii.2, p. 12.

45   Ibid., I.ii.21–23, pp. 21–22; IV.ii.1, p. 433. It is true that Locke says that "It is on this intuition that depends all the certainty and evidence of all our knowledge, which certainty every one finds to be so great that he cannot imagine, and therefore not require, a greater" (p. 433). Yet the fact remains that this is a certainty about *ideas*, and not *the world*.

46   Ibid., IV.iii.29, p. 456.

47 George Berkeley, *A New Theory of Vision and other Select Philosophical Writings* (London: J.M. Dent and E.P. Dutton, 1914), 201.

48 Ibid., 246.

49 Ibid., 265.

50 Ibid., 271.

51 Thomas Reid, *Essays on the Intellectual Powers of Man*, ed. Derek R. Brookes (University Park: Pennsylvania State University Press, 2002), 464.

52 Berkeley, *A New Theory of Vision*, 259.

53 The list includes Galileo, Descartes, Gassendi, and Hobbes. For Galileo see *The Assayer*, in *Discoveries and Opinions of Galileo*, 274–78. For Descartes see *Principles of Philosophy*, in *The Philosophical Writings of Descartes Vol. 1*, trans. John Cottingham, Robert Stoothoff, Dugald Murdoch (Cambridge: Cambridge University Press, 1985), 282–85; *Meditations*, in *The Philosophical Writings of Descartes Vol. 2*, trans. John Cottingham, Robert Stoothoff, Dugald Murdoch (Cambridge: Cambridge University Press, 1985), 30. For Gassendi see Popkin, *The History of Scepticism from Erasmus to Spinoza*, 100–103. For Hobbes see *Leviathan*, I, i, 85–87.

54 Locke, *An Essay Concerning Human Understanding*, II.viii.10, p. 85.

55 Ibid., II.viii.15, p. 87.

56 See Pierre Bayle, "Pyrrho," in *Pierre Bayle, Historical and Critical Dictionary: Selections*, ed. and trans. Richard H. Popkin (Indianapolis: Bobbs-Merrill, 1965), 197–98.

57 Berkeley, *A New Theory of Vision*, 259–60.

58 Ibid., 273.

59 Ibid., 200.

60 Reid, *Intellectual Powers*, 141, 148.

61 Berkeley, *A New Theory of Vision*, 241.

62 For a recent, wide-ranging treatment of the rise of "common sense" as a term of modern democratic politics, see Sophia Rosenfeld, *Common Sense: A Political History* (Cambridge & London: Harvard University Press, 2011).

63 Kathleen Wilson, *The Sense of the People: Politics, Culture and Imperialism in England, 1715–1785* (Cambridge: Cambridge University Press, 1995). "The claim to represent the 'sense of the people' became an important legitimizing rhetorical strategy in the Hanoverian decades, a crucial part of the wider political contestation under way that had been produced by the emergence of a vibrant, national and predominantly urban extraparliamentary political culture." Ibid., 3.

64 See David F. Norton, *David Hume: Common-Sense Moralist, Skeptical Metaphysician* (Princeton: Princeton University Press, 1982); David F. Norton, ed., *The Cambridge Companion to Hume* (Cambridge: Cambridge University Press, 1993). Kenneth Clatterbaugh argues that Hume tried, but failed, to provide a positive account of causation that brought philosophy into alignment with modern science. See Clatterbaugh, *The Causation Debate*, Chapter 8.

65 David Hume, *A Treatise of Human Nature*, ed. L. A. Selby-Bigge (Oxford: Clarendon Press, 1978), 263–74. In one rhetorical flight of fancy, Hume claims to be "affrighted and confounded with that forelorn solitude, in which I am plac'd in my philosophy, and fancy myself some strange uncouth monster, who not being able to mingle and unite with society, has been expell'd all human commerce, and left utterly abandon'd and disconsolate." Ibid., 263. While Hume was in fact a very sociable and well-liked character, he was clearly aware of the degree to which his ideas took him away from the perceptions of "common life" (267).

66   Ibid., 268–69. See also David Hume, *An Enquiry Concerning Human Understanding*, ed.
     Tom L. Beauchamp (Oxford: Oxford University Press, 1999), 120.
67   David Hume, *Essays: Moral, Political, and Literary*, ed. Eugene F. Miller
     (Indianapolis: Liberty Classics, 1987), 111; Hume, *An Enquiry concerning Human
     Understanding*, 88, 175, 204, 207. Hume took the existence of causal relations in nature
     for granted. The main thrust of his discussion of causality was to question its *intel-
     ligibility*. See Galen Strawson, *The Secret Connexion: Causation, Realism and David Hume*
     (Oxford: Clarendon Press, 1989).
68   Norton, *David Hume*, 43, 109, 135.
69   David Hume, *An Abstract of A Treatise on Human Nature*, ed. J. M. Keynes and P. Sraffa
     (Cambridge: Cambridge University Press, 1938), 5–6.
70   Hume, *A Treatise of Human Nature*, 193, 267, 150, 271. In *An Enquiry Concerning Human
     Understanding*, Hume stated that "the generality of mankind" supposes that "they per-
     ceive the very force or energy of the cause" of observed events, yet "philosophers, who
     carry their scrutiny a little farther, immediately perceive, that, even in the most familiar
     events, the energy of the cause is as unintelligible as in the most unusual, and that we
     only learn by experience the frequent CONJUNCTION of objects, without being ever
     able to comprehend any thing like CONNEXION between them" (140–41).
71   Hume, "Of Commerce," in *Essays: Moral, Political, and Literary*, ed. Eugene F. Miller
     (Indianapolis: Liberty Classics, 1987), 253.
72   Hume, "The Rise of the Arts and Sciences," in *Essays: Moral, Political, and Literary*, 120.
73   Harvey Chisick, "David Hume and the Common People," in *The "Science of
     Man" in the Scottish Enlightenment: Hume, Reid and Their Contemporaries*, ed. Peter
     Jones (Edinburgh: Edinburgh University Press, 1989), 12. See also John Dwyer,
     "Introduction—A 'Peculiar Blessing': Social Converse in Scotland from Hutcheson
     to Burns," in *Sociability and Society in Eighteenth-Century Scotland*, ed. John Dwyer and
     Richard B. Sher (Edinburgh: Mercat Press, 1993), 1–20.
74   Donald W. Livingston, *Hume's Philosophy of Common Life* (Chicago: University of
     Chicago Press, 1984), 30.
75   See Hume's discussion of his "science of man" in *A Treatise of Human Nature*,
     xv–xix, 273.

Chapter 3 Common Notions, *Sens Commun*: Herbert of Cherbury and Renè Descartes

1   Thomas Reid, *An Enquiry into the Human Mind, on the Principles of Common Sense*, ed.
     Derek R. Brookes (University Park: Penn State University Press, 1997), 38.
2   As will become evident (see Chapter 8), Reid was actually quite aware of his
     predecessors in the common-sense philosophical tradition and was happy to cite them
     when it suited him.
3   Henry Lee used the term in the same way that Reid did, in his *Anti-Scepticism: Or, Notes
     Upon Each Chapter of Mr. Lock's Essay Concerning Humane Understanding* (London: Clavel &
     Harper, 1702), 272. Discussed further in Chapter 5.
4   Discussed further in Chapter 7. There were many facets to this movement. My use
     of the term is broad and includes figures like Hobbes, Descartes, Mandeville, and
     Hume, all of whom applied conceptions in line with one or another aspect of the

New Science to the study of human nature and/or society. Thomas Reid was clearly a part of this movement, and his self-understanding as an empirical student of the human mind will be discussed further in Chapter 8.

5   See Richard H. Popkin, *The History of Scepticism from Erasmus to Spinoza* (Berkeley: University of California Press, 1979), Chapter III.

6   Ibid., 52–53.

7   See Popkin, *The History of Scepticism*, Chapters XI and XII; Richard Popkin, "Bayle and the Enlightenment" and "Hume" in *The Encyclopedia of Philosophy, Vol. 7* (New York: Macmillan, 1967); Walter Rex, *Essays on Pierre Bayle and Religious Controversy* (The Hague: Martinus Nijhoff, 1965); Elisabeth Labrousse, *Bayle* (Oxford: Oxford University Press, 1983).

8   Popkin, *The History of Scepticism*, 53.

9   Michel de Montaigne, *The Complete Essays*, trans. M. A. Screech (New York: Penguin Books, 1991), 679.

10 Ibid., 642–43.

11 For extended discussion of the mitigated skepticism of Mersenne and Gassendi, see Popkin, *The History of Scepticism*, Chapter VII.

12 Montaigne, *The Complete Essays*, 647.

13 This trope is especially evident in his essay "On the Cannibals." See Montaigne, *The Complete Essays*, 228–41.

14 John Locke, *An Essay Concerning Human Understanding*, 6th ed. (Amherst: Prometheus Books, 1995), I.iii, pp. 26–32; quote from p. 31.

15 See Daniel Carey, "Locke as Moral Sceptic: Innateness, Diversity, and the Reply to Stoicism," *Archive für Geschichte der Philosophie*, 79 (1997): 292–309.

16 Locke, *An Essay Concerning Human Understanding*, "The Epistle to the Reader," xvi.

17 Edward Stillingfleet, *The Bishop of Worcester's Answer to Mr. Locke's Letter, Concerning some Passages Relating to His Essay of Humane Understanding* (London: Henry Mortlock, 1697), 90.

18 La Mothe le Vayer, "Opuscule ou Petit Traité Sceptique, sur cette commun Facon de Parler, N'AVoir Pas les Sens-commun" (1646), in *L'esprit de La Motte le Vayer*, ed. Charles Antoine Leclerc de Montlinot (1763), 364.

19 Ibid., 369.

20 Ibid., 365–66.

21 For more on the diversity problem see Daniel Carey, "Method, Moral Sense, and the Problem of Diversity: Frances Hutcheson and the Scottish Enlightenment," *British Journal for the History of Philosophy*, 5, no. 2 (1997): 275–96.

22 See Locke, *An Essay Concerning Human Understanding*, I.iii.15–20, pp. 35–39.

23 For discussion of the later addition of the chapter on religious Common Notions see the "Introduction," in *De veritate, by Edward, Lord Herbert of Cherbury*, trans. Meyrick H. Carré (Bristol: University of Bristol, 1937), 15, 55.

24 For more on Herbert see R. D. Bedford's excellent study, *The Defence of Truth: Herbert of Cherbury and the Seventeenth Century* (Manchester: Manchester University Press, 1979).

25 Quoted in ibid., 4.

26 As Bedford aptly puts it, "Behind the philosopher sharpening his categories lurks the duellist sharpening his sword [...] Herbert, hungry for distinction and the applause of others, took his chivalric vows as seriously in their way as he took the defence of distressed Truth." Bedford, *The Defence of Truth*, 11.

27 *De Veritate*, 106. For discussion of Stoic "common notions" see E. Vernon Arnold, *Roman Stoicism* (London: Routledge & Kegan Paul, 1958), 136–43.
28 "By faculty I mean every inner power which develops the different forms of apprehension in their relation to the different forms of the objects." *De Veritate*, 108.
29 Herbert's thought is infused with the widespread Renaissance notion, derived from a number of sources including Plato and Neo-Platonism, Paracelsus, and the Hermetic writings, of an analogy or correspondence between man and the universe. See the discussion in Bedford, *The Defence of Truth*, Chapter IV. For examples of Herbert's employment of the microcosm/macrocosm analogy, see *De Veritate*, 91, 108, 133, 189, 207.
30 For Herbert, reason, understood as a discursive faculty that makes logical inferences, is clearly subordinate to instinctual first principles or Common Notions. It draws upon Common Notions in ascertaining other truths, but rests ultimately on the primary Common Notions. Left unconstrained by the truths of Natural Instinct, "reason remains undecided and will raise such a host of images and scruples of its own that it will never reach any conclusion." *De Veritate*, 141; see also 116, 120–22, 131, and 137–39.
31 *De Veritate*, 78.
32 Ibid., 80.
33 Ibid., 90.
34 Ibid., 91–104.
35 Ibid., 115.
36 Ibid., 88.
37 Ibid., 105.
38 Ibid.
39 Ibid.
40 Ibid., 87.
41 Ibid., 126.
42 Ibid., 130; 107.
43 Ibid., 133.
44 Ibid., 296.
45 Ibid., 289–308.
46 Ibid., 314.
47 Ibid., 135; see also 126.
48 Ibid., 136. Herbert's friendship with the natural law theorist Hugo Grotius likely played a role here. See the discussion in Bedford, *The Defence of Truth*, 214–15.
49 *De Veritate*, 184–86.
50 Ibid., 185–89.
51 Ibid., 139–40.
52 Ibid., 117.
53 Locke, *An Essay Concerning Human Understanding*, I.iii.19–21, pp. 38–39; Meyrick H. Carré, "Introduction", in *De Veritate*, 34, 65.
54 Herbert in fact argues that popular errors spread widely when they contain a large admixture of truth and that "the more sublime and essential truths are, the more they are liable to be mixed with error." Ibid., 82; see also 130. For the Stoic view of "general consent" see E. Vernon Arnold, *Roman Stoicism* (London: Routledge & Kegan Paul, 1958), 143; 223–24; 325.
55 *De Veritate*, 117.

56 Ibid., 140. Herbert says elsewhere that "Common Notions cannot be denied except by madmen." Ibid., 116. For other examples of statements by Herbert about the spontaneous and ineluctable conviction produced by instinctual Common Notions, see ibid., 128–30, 135.
57 Ibid., 78–79.
58 Ibid., 118.
59 See the discussion in ibid., Chapter IX.
60 Quoted in Bedford, *The Defence of Truth*, 3.
61 *De Veritate*, 58–59, 120, 137, 162–65.
62 Ibid., 303.
63 Ibid., 72.
64 Ibid., 118.
65 One of the few explicit references Herbert makes to previous thinkers is to Cicero: "As the Prince of Orators has well said: 'God has enclosed intelligence in the mind, and the mind in the body.'" Ibid., 86. For discussion of the many plausible sources of Herbert's ideas see Bedford, *The Defence of Truth*, passim and Chap. IV. See also Meyrick Carré's Introduction to *De Veritate*, esp. 17–18, 32–33, 42n2, 60.
66 Ibid., 106.
67 See, for example, his statement to this effect in ibid., 133.
68 See Bedford, *The Defence of Truth*, esp. Chapters V–VIII. See also Carré's comments in *De Veritate*, 58–62.
69 Sir William Hamilton states that *De Veritate* "contain[s] a more formal and articulate enouncement of the doctrine of common sense, than had (I might almost say than has) hitherto appeared." *The Works of Thomas Reid Vol. 2*, ed. Sir William Hamilton (Edinburgh: MacLachlan and Stewart, 1863), 781. Charles de Rémusat published a book on Herbert titled *Lord Herbert de Cherbury, sa vie et ses oeuvres, ou les origines de la philosophie du sens commun et de la théologie naturelle en Angleterre* (Paris, 1874).
70 Quoted in *De Veritate*, 24–25.
71 Bedford, *The Defence of Truth*, 55–57.
72 Quoted in ibid., 57.
73 See Popkin, *The History of Skepticism*, Chapter IX.
74 *The Philosophical Writings of Descartes vol. 1*, trans. John Cottingham, Robert Stoothoff, and Dugald Murdoch (Cambridge: Cambridge University Press, 1999), 9, 111, 117; vol. 2, 409, 415. Quote is from vol. 1, 111.
75 *The Philosophical Writings of Descartes vol. 1*, 117.
76 Ibid., 32.
77 Ibid., 48.
78 Ibid., 111.
79 Ibid., 121.
80 Ibid., 34.
81 Ibid., 35.
82 *The Philosophical Writings of Descartes vol. 2*, 405.
83 *The Philosophical Writings of Descartes vol. 1*, 150.
84 Ibid., 355; *The Philosophical Writings of Descartes vol. 2*, 412.
85 *Sens commun* was first used extensively as a philosophical term in France in the early eighteenth century, in Buffier's writings, and the entry in the *Encylopédie* of Diderot and D'Alembert on *Sens Commun* was taken directly from his *Traité des Premieres Veritez* (discussed further in Chapter 6).

86 See J. Nicot, *Thresor de la langue Francoyse* (Paris: D. Douceur, 1606); César Oudin, *Le Thresor des trois langues, espagnole, françoise, et italienne* (Genève: J. Crespin, 1627); Guy Miege, *The Great French dictionary[…] English and French* (London: Thomas Basset, 1688); *Dictionaire universel* (La Haye; Rotterdam: Arnout et R. Leers, 1690).

87 *Nouveau dictionnaire de l'Académe françoise* (Paris: J.B. Coignard, 1718).

88 Here *sens commun* is defined in part as "those general concepts or ideas which are born in the spirit/mind (*esprit*) of all men; certain natural lights which allow them to judge things in the same way." *Dictionnaire universel françois et latin* (Paris: Trevoux, 1721).

89 Quoted in *The Encyclopedia of Diderot & d'Alembert Collaborative Translation Project*, http://www.hti.umich.edu/d/did/, retrieved 2/7/2005. From vol. 2, page 328, of the *Encyclopédie*.

90 *The Philosophical Writings of Descartes vol. 1*, 48.

91 Ibid., 49.

92 See ibid., 22, 44–48, 224, 303–4; vol. 2, 26–27.

93 *The Philosophical Writings of Descartes vol. 1*, 304. See also ibid., 44–45, for a list of other "simple natures which the intellect recognizes by means of a sort of innate light."

94 *The Philosophical Writings of Descartes vol. 2*, 52.

95 *The Philosophical Writings of Descartes vol. 1*, 224. See also *vol 2*, 53.

96 *The Philosophical Writings of Descartes vol. 2*, 57.

97 *The Philosophical Writings of Descartes vol. 1*, 42. See also vol. 1, 164, 303.

98 See the *Treatise on Man*, in *The Philosophical Writings of Descartes vol. 1*, 105–8; and *The Passions of the Soul*, in ibid., 340–3. Quote from 106.

99 *The Philosophical Writings of Descartes vol. 1*, 195. See also *vol. 2*, 417–18.

100 See *The Philosophical Writings of Descartes vol. 1*, 194, 301; *vol. 2*, 55. While Descartes does assert in the *Principles of Philosophy* that "we have a clear and distinct perception of some kind of matter" (*vol. 1*, 223), he immediately follows this assertion by the same argument presented in the *Meditations* that such perceptions cannot be deceiving because "it is quite inconsistent with the nature of God that he should be a deceiver."

Chapter 4 Hobbes, Locke, and Innatist Responses to Skepticism and Materialism

1   Hobbes met Galileo and corresponded with Descartes and also personally knew Mersenne, Gassendi, and Bacon. While scholars don't always agree on the relative impact of these thinkers on Hobbes, all agree that Galileo and Descartes were especially important in the genesis of Hobbes's mature thought. See, for example, Richard Tuck, *Hobbes* (Oxford: Oxford University Press, 1992).

2   Thomas Hobbes, *Leviathan* (New York: Penguin Classics, 1985), I.i, p. 85.

3   Ibid., I.i, pp. 85–86.

4   Ibid., I.i, p. 86.

5   Ibid., I.iii, pp. 98–99.

6   Ibid., I.v, p. 115.

7   Ibid., I.vii, pp. 133–34.

8   Ibid., I.vi, p. 120.

9   Ibid., I.iv, p. 105.

10 Many people assumed Hobbes was an atheist, but some modern scholars are not so sure. See Patricia Springborg, "Hobbes on Religion," in *The Cambridge Companion to Hobbes*, ed. Tom Sorrell (Cambridge: Cambridge University Press, 1996), 346–80.

11   See David Berman, *A History of Atheism in Britain: From Hobbes to Russell* (London: Croom Helm, 1988), Chapter 2; Springborg, "Hobbes on Religion," 347–48; Samuel I. Mintz, *The Hunting of Leviathan* (Cambridge: Cambridge University Press, 1962); A. P. Martinich, *Hobbes: A Biography* (Cambridge: Cambridge University Press, 1999), Chapter 10.

12   Berman, *A History of Atheism in Britain*, 48.

13   Ibid., 57–59; Martinich, *Hobbes*, 329–31.

14   John Yolton, *John Locke and the Way of Ideas* (Oxford: Oxford University Press, 1956), 29.

15   Ibid., 31.

16   Ibid., 31–33.

17   Quoted in ibid., 35–36.

18   Quoted in ibid., 33.

19   For affinities between Herbert and the Cambridge Platonists see R. D. Bedford, *The Defence of Truth: Herbert of Cherbury and the Seventeenth Century* (Manchester: Manchester University Press, 1979), 70–71, 90–91. According to Carré, although many passages in the writings of these individuals echoed the thought of Herbert, Nathaniel Culverwel was the only member of this school explicitly to discuss Herbert's theory, "on many points of which he is in agreement." See *De Veritate, by Edward, Lord Herbert of Cherbury*, trans. Meyrick H. Carré (Bristol: University of Bristol, 1937), 43. The issue here, as elsewhere, is less one of direct influence than of the existence of a general intellectual climate in which such ideas were developed by a variety of thinkers, eventually coalescing into an identifiable tradition of modern thought.

20   Henry More, *A Collection of Several Philosophical Writings of Dr. Henry More, vol. 1* (London: James Fleisher, 1662), 7.

21   Ibid., 13, 29.

22   Ibid., 17.

23   Ibid., 17–18. More suggests that the term "innate idea" is itself expendable, but he thinks that it helps to clarify what he is talking about. See the appendix in ibid., 147–48.

24   Ibid., 17.

25   Ibid., 18–19.

26   Ibid., 70.

27   Ibid., 65–67.

28   Ibid., 69.

29   Ibid., 98.

30   Henry More, *A Collection of Several Philosophical Writings of Dr. Henry More, vol. 2* (London: James Fleisher, 1662), 101–9.

31   More, *A Collection of Several Philosophical Writings of Dr. Henry More, vol. 1*, 97–98.

32   John Locke, *An Essay Concerning Human Understanding*, 6th ed. (Amherst: Prometheus Books, 1995), I.iii.27, p. 41.

33   Ibid., I.iii.22, p. 39.

34   Quoted in G. A. J. Rogers, "Locke, Newton, and the Cambridge Platonists on Innate Ideas," in *Philosophy, Religion and Science in the Seventeenth and Eighteenth Centuries*, ed. John W. Yolton (Rochester: University of Rochester Press, 1990), 193–94. See also the discussion in Daniel Carey, "Locke as Moral Sceptic: Innateness, Diversity, and the Reply to Stoicism," *Archiv für Geschicte der Philosophie*, 79 (1997): 302–3; and Yolton, *John Locke and the Way of Ideas*, 44–66.

35   Locke, *An Essay Concerning Human Understanding*, I.i.Introduction. See also Yolton, *John Locke and the Way of Ideas*, esp. Chapter II.

36   Thomas Reid, *Essays on the Intellectual Powers of Man*, ed. Derek R. Brookes (University Park: Pennsylvania State University Press, 2002), 433.

37   Robert Ferguson, *The Interest of Reason in Religion: With the Import & Use of Scripture-Metaphors, and the Nature of the Union Betwixt Christ & Believers* (London: Printed for Dorman Newman, 1675), 22.

38   Ibid., 22–23.

39   Ibid., 23–24.

40   Discussed further in Chapter 8. Reid's "acquired perceptions" are much more clearly the result of experience than Ferguson's "acquired principles," which are described more as the result of rational deduction from first principles. Nevertheless, Ferguson clearly assumes experience to be part of the process of ascertaining acquired principles.

41   Ferguson, *The Interest of Reason in Religion*, 24.

42   Ibid., 27; 19.

43   Ibid., 41.

44   Ibid., 40.

45   Ibid., 237.

46   Ibid., 234–36.

47   Indeed, a central theme of Rosenfeld's study of the political history of common sense is the malleability and multi-valency of the term, in social and political discourse. See Sophia Rosenfeld, *Common Sense: A Political History* (Cambridge: Harvard University Press, 2011).

48   Ferguson, *The interest of reason in religion*, 42–43. The idea of a natural moral virtue, as opposed to religious virtue, had been taken up in Ferguson's *A Sober Enquiry into the Nature, Measure, and Principle of Moral Virtue, [and] Its Distinction from Gospel-Holiness* in 1673.

49   For more see Yolton, *John Locke and the Way of Ideas*. It is beyond the scope of this book to examine all the varieties of seventeenth-century innatism or to chronicle every early use of the term "common sense" in philosophical discourse (and I suspect it is beyond the patience of the reader as well). Yolton exhaustively analyzes the many varieties of innatism in English thought, before chronicling the various critiques of Locke's "way of ideas." Yolton makes clear that Locke was responding to an entrenched epistemology that, as in the case of Ferguson, began to recognize "common sense" as an analogue of "natural reason." Yolton's discussion furthermore demonstrates that seventeenth-century appeals to innate ideas or principles of mind, in defense of established religion, paved the way for the more naturalistic and less theological discussions of moral sense and common sense that followed in the eighteenth century.

50   Ferguson, *The Interest of Reason in Religion*, 240.

51   Ibid., 237.

52   The headings "common" and "sense" do not contain any reference to "common sense" in the following dictionaries: Claudius Hollyband, *A dictionarie French and English* (London: Thomas Woodcock, 1593); Randle Cotgrave, *A Dictionarie of the French and English Tongues* (London: Adam Islip, 1611); Randle Cotgrave, *A French-English dictionary* (London: George Lathum, 1650); John Rider, *Riders dictionary* (London: Andrew Cook, 1659); James Howell, *Lexicon Tetraglotton: An English-Italian-Spanish dictionary* (London: Thomas Leach, 1660); Francis Gouldman, *A Copious Dictionary in Three Parts* (Cambridge: John Hayes, 1674); Edward Phillips, *The New World of Words, or, a General English Dictionary* (London: Obadiah Blagrave, 1678); Guy Miege, *The Great French Dictionary … English and French* (London: Thomas Basset, 1688).

53 See *The Oxford English Dictionary vol. III* (Oxford: Clarendon Press, 1989).

54 For example: "Cela est bon sens à eux. *That is wisely done of them, or turnes greatly to their advantage.*" Randle Cotgrave, *A Dictionarie of the French and English Tongues* (London: Adam Islip, 1611); Repeated in Randle Cotgrave, *A French-English Dictionary* (London: George Lathum, 1650).

55 Guy Miege, *A New Dictionary French and English* (London: Tho. Dawks, 1677); Guy Miege, *The Great French Dictionary … English and French* (London: Thomas Basset, 1688).

56 Quoted in Rosenfeld, *Common Sense*, 264.

57 Samuel Johnson, *A Dictionary of the English Language* (London: W. Strahan, 1755).

58 Locke, *An Essay Concerning Human Understanding*, I.iii.27, p. 41.

59 Ibid., I.ii.1, p. 12.

60 Ibid., I.ii.3, p. 12; I.iii.19, p. 38.

61 Ibid., I.ii.5, p. 13.

62 See the discussion in ibid., I.iii.1–19, pp. 26–38.

63 Ibid., I.iv.10–11, p. 46.

64 Locke, *An Essay Concerning Human Understanding*, I.iv.12, p. 47. See also IV.x.1–6, pp. 527–29. Here Locke states that "sense, perception, and reason" are all that is needed to provide us with a clear proof of God's existence.

65 Ibid., I.iii.3, p. 28.

66 For a clear and concise statement by Locke of this view see the discussion in Book II, "Of Complex Ideas," in ibid., II.xii.1, p. 108.

67 Continuing, Locke states: "For in this the mind is at no pains of proving or examining, but perceives the truth, as the eye doth light, only by being directed towards it. Thus the mind perceives that white is not black, that a circle is not a triangle, that three are more than two, and equal to one and two." Such knowledge is "irresistible, and, like bright sunshine, forces itself immediately to be perceived as soon as ever the mind turns its view that way [...] It is on this intuition that depends all the certainty and evidence of all our knowledge." Ibid., IV.ii.1, p. 433.

68 Ibid., IV.vii.2, p. 505.

69 Ibid., I.ii.18–26, pp. 19–24.

70 Ibid., II.xxvi.1, p. 238. It should be noted that Locke does not, in the *Essay*, refer to causality as a "self-evident principle." His discussion of self-evident principles is taken up with the principles of identity and noncontradiction, since he believes that they have the most reason to be called innate (I.ii.4, p. 13). Yet Locke's discussion of self-evident principles would seem to include all principles that become obvious once the terms are understood, including causality (II.xxvi.2, p. 239). The question to ask is how we are able to discern a causal relation from two discrete events, if we don't already have this idea in mind and hence *expect* for there to be such relations in nature. As it happens, in his discussion "Of Power," Locke states that the mind "*must* [italics mine] collect a power somewhere, able to make [an observed] change" in sensible qualities (II.xxi.4, p. 165), suggesting that we are indeed compelled to look for causal connections in nature. Locke then goes on to suggest that our idea of active power (or cause) may be the product of reflection on our own ability to will movements of our body. However, this does not solve the problem of how we are able to make a firm causal connection between our acts of mind and sundry bodily motions, unless we are able to intuit some such relation between any two experiences. Hume would later argue that we just get into the *habit* of making such connections, but Locke was too much of a rationalist and a theist to consider such a notion.

71   Ibid., I.iv.25, p. 56.
72   For more on the Royal Society and the methods of the New Science in general I refer
      the reader to the discussion and citations made in Chapter 2. For substantiation of this
      point from the *Essay*, see especially IV.xii.5–10, pp. 546–48.
73   Locke, *An Essay Concerning Human Understanding*, I.iii.4, p. 28.
74   "Nobody will so openly bid defiance to common sense as to affirm visible and direct
      contradictions in plain words." Ibid., IV.viii.2, p. 520.
75   Ibid., II.xxxiii.18, p. 320; IV.xix.7, p. 591. Quote from IV.xviii.11, p. 589. The full
      passage reads:

> For men, having been principled with an opinion that they must not consult
> reason in the things of religion, however apparently contradictory to common
> sense and the very principles of all their knowledge, have let loose their fancies
> and natural superstition; and have been by them led into so strange opinions
> and extravagant practices in religion, that a considerate man cannot but stand
> amazed at their follies.

76   Ibid., I.iii.26, p. 41.
77   Ibid., IV.x.3, p. 528. See also Locke's discussion of intuitive knowledge in IV.ii.1,
      p. 433.
78   For more see Yolton, *John Locke and the Way of Ideas*, esp. 24–25, 64–71.
79   Ephraim Chambers, "Notions," in *Cyclopædia: Or, an Universal Dictionary of Arts and
      Sciences, vol. 2* (London: James and John Knapton, 1728). The first quotation is from
      the entry "Innate Ideas" and the second is from the entry "Notions."

Chapter 5 Common Sense in Early Eighteenth-Century Thought

1   Robert Ferguson, *The Interest of Reason in Religion: With the Import & Use of Scripture-
     Metaphors, and the Nature of the Union Betwixt Christ & Believers* (London: Printed for Dorman
     Newman, 1675), 235–36; and John Ray, *The Wisdom of God Manifested in the Works of the
     Creation, in Two Parts* (London: Samuel Smith, Prince's Arms, St. Paul's, 1692), Part 1, 43.
2   Sophia Rosenfeld, *Common Sense: A Political History* (Cambridge: Harvard University
     Press, 2011).
3   John Yolton, *John Locke and the Way of Ideas* (Oxford: Oxford University Press, 1956), 14.
4   For extensive discussion and analysis of the response to Locke's *Essay*, see ibid.,
     Chaps. I–III.
5   John Sergeant, *Solid Philosophy Asserted, against the Fancies of the Ideists, or, The Method to Science
     Farther Illustrated* (London: Roger Clavil, Abel Roper, Thomas Metcalf, 1697), 24, 49.
6   Edward Stillingfleet, *The Bishop of Worcester's Answer to Mr. Locke's Second Letter, Wherein
     His Notion of Ideas Is Prov'd to Be Inconsistent with It Self, and with the Articles of the Christian
     Faith* (London: Henry Mortlock, 1698), 71; see also 25, 54–55.
7   Edward Stillingfleet, *The Bishop of Worcester's Answer to Mr. Locke's Letter, Concerning Some
     Passages Relating to His Essay of Human Understanding* (London: Henry Mortlock, 1697), 96.
8   Stillingfleet, *The Bishop of Worcester's Answer to Mr. Locke's Second Letter*, 28–30.
9   Ibid., 146; Stillingfleet, *The Bishop of Worcester's Answer to Mr. Locke's Letter*, 95. In both
     texts Stillingfleet equates "principles of reason" with self-evident perceptions like
     "everything we see must have a cause" (ibid., 94).

10 E.g. Stillingfleet, *The Bishop of Worcester's Answer to Mr. Locke's Letter*, 90; and Stillingfleet, *The Bishop of Worcester's Answer to Mr. Locke's Second Letter*, 54–55, 71.

11 Henry Lee, *Anti-Scepticism: Or, Notes Upon Each Chapter of Mr. Lock's Essay Concerning Humane Understanding* (London: Clavel & Harper, 1702). Quote from *The Dictionary of Seventeenth-Century British Philosophers, vol. 2*, ed. Andrew Pyle (Bristol, England: Thoemmes Press, 2000), 507.

12 Lee, *Anti-Scepticism*, Epistle Dedicatory.

13 Ibid., 27. For other examples see the preface, 31 and 272.

14 Ibid., 27.

15 Ibid., preface.

16 Ibid.

17 Ibid., 7.

18 Ibid., 36.

19 Ibid., 14.

20 Ibid., 13.

21 Ibid., 11. The gist of Lee's arguments presented here turn out to be well-supported by modern research—discussed in the Epilogue.

22 Ibid., 267.

23 Ibid., 267–68.

24 Ibid., 268.

25 Ibid., 29.

26 Ibid., 26–27.

27 Ibid., 272.

28 As, for example, when he exclaims, in reference to Locke's notion that we cannot have knowledge of first truths if we do not have a precise understanding of the words that describe them, "Is not all this contrary to common sense?"; or when he states that Hobbes' supposition that all constituents of the universe are reducible to matter and motion is "contrary to common sense and reason." Ibid., 27, 31.

29 As with terms like "the public," "common sense" never achieved the same resonance in the socially hierarchical and politically fragmented German states as it did in Britain and France, although populist philosophers like J. G. Herder did begin to toy with the term beginning in the 1760s. For more see Benjamin W. Redekop, *Enlightenment and Community: Lessing, Abbt, Herder, and the Quest for a German Public* (Montreal: McGill-Queen's University Press, 2000).

30 G. W. Leibniz, *New Essays on Human Understanding*, trans. and ed. Peter Remnant and Jonathan Bennet (Cambridge: Cambridge University Press, 1996), 80.

31 Ibid.

32 Ibid., 375–92.

33 Ibid., 49–52, 74–76, 86–91.

34 Ibid., 50.

35 Ibid., 50, 51, 76, 90, 393.

36 Ibid., 52.

37 Ibid., 84.

38 As he says, "It would indeed be wrong to think that we can easily read these eternal laws of reason in the soul [...] without effort or inquiry." Ibid., 50; see also 74–76, 107.

39 Ibid., 50, 84.

40 Just as geometers have been satisfied to deduce a great many theorems from a few rational principles, so it will be enough "if practitioners of natural science can, by

means of certain principles of experience, account for a great many phenomena and even predict them in practice." Ibid., 453.

41   Ibid., 455.

42   See Peter Earle, *The Making of the English Middle Class: Business, Society and Family Life in London, 1660–1730* (Berkeley: University of California Press, 1989); Kathleen Wilson, *The Sense of the People: Politics, Culture and Imperialism in England, 1715–1785* (Cambridge: Cambridge University Press, 1995). Rosenfeld, *Common Sense*, Chapter 1.

43   Rosenfeld, *Common Sense*, 30.

44   *The Tatler*, No. 111, Saturday, December 24, 1709.

45   *The Spectator*, No. 234, Wednesday, November 28, 1711.

46   *The Tatler*, No. 158, Thursday, April 30, 1710.

47   *The Spectator*, No. 70, Monday, May 21, 1711.

48   *The Spectator*, No. 105, Saturday, June 30, 1711.

49   *The Spectator*, No. 29, Tuesday, April 3, 1711.

50   Hogarth argued that aesthetic canons should be rooted in "the familiar path of common observation [...] Even a butcher or blacksmith can show himself to be a critic in proportion," and ladies "will often point out such particular beauties or defects in [the make of necks, hands and arms] as might easily escape the observation of a man of science." See William Hogarth, *The Analysis of Beauty: Written with a View of Fixing the Fluctuating Ideas of Taste* (London: J. Reeves, 1753), 79–80.

51   Alexander Pope, *An Essay on Criticism*, in *The Poetical Works of Alexander Pope*, ed. H. F. Cary (New York: D. Appleton, 1867), 45. Pope in fact often appealed to "good sense" or "common sense" in a similar fashion, as an inborn power of judgment rooted in nature, as when he states in his *Moral Essays* (1731–35) that good sense is the first principle and foundation of taste as well as everything else, "The chief proof of [which] is to *follow nature*, even in works of mere luxury and elegance." Alexander Pope, *Moral Essays*, in *The Poetical Works of Alexander Pope*, 279; see also 4, 57, 281. In the Prolegomena to *The Dunciad* (1728) Pope writes: "Let others creep by timid steps, and slow, / On plain experience lay foundations low, / By common sense to common knowledge bred, And last, to Nature's cause through Nature led." Alexander Pope, *The Dunciad*, in *The Poetical Works of Alexander Pope*, 460; see also 241, 382, 415.

52   Addison writes in early 1711 that care for our health is prompted "not only by common sense, but by duty and instinct" (*The Spectator*, No. 25, March 29, 1711), and later that year he chides hens for having all the instincts needed to lay and hatch eggs, yet being "without the least glimmering of thought or common sense" (*The Spectator*, No. 120, Wednesday, July 18, 1711).

53   *The Spectator*, No. 259, Thursday, December 27, 1711.

54   *The Spectator*, No. 64, Monday, May 14, 1711.

55   *The Spectator*, No. 156, Wednesday, August 29, 1711; *The Spectator*, No. 126, Wednesday, July 25, 1711 (Joseph Addison).

56   For Shaftesbury's impact on German thinkers see Redekop, *Enlightenment and Community*, 86–88, 129–32, 142, 155. See also John Andrew Bernstein, "Shaftesbury's Optimism and Eighteenth-Century Social Thought," in *Anticipations of the Enlightenment in England, France, and Germany*, ed. Alan C. Kors and Paul J. Korshin (Philadelphia: University of Pennsylvania Press, 1987), 87–94. Shaftesbury's influence on the subsequent "moral sense" school of British thought, which formed part of the background to the rise of the philosophy of common sense, is well known and discussed further in Chapter 6. For more on Shaftesbury and the moral sense school see David Fate Norton, *From Moral*

Sense to Common Sense: An Essay on the Development of Scottish Common Sense Philosophy, 1700–1765 (PhD Dissertation, University of California at San Diego, 1966).

57   See especially "An Inquiry Concerning Virtue, or Merit," in *Characteristicks Vol. II*, ed. Anthony Ashley Cooper, Earl of Shaftesbury (London, 1732); quote from p. 90. See also Stanley Grean, *Shaftesbury's Philosophy of Religion and Ethics: A Study in Enthusiasm* (Athens: Ohio University Press, 1967), esp. Chaps. 9–10; Norton, *From Moral Sense to Common Sense*, Chap. III.

58   Anthony Ashley Cooper, Earl of Shaftesbury, "Sensus Communis: An Essay on the Freedom of Wit and Humour," *Characteristicks Vol. I* (London, 1732), 104; see also 111.

59   Ibid., 79–80.

60   Ibid., 80. As Shaftesbury put it in a letter to Michael Ainsworth, Locke completed what Hobbes began: "'Twas Mr. Locke that struck at all fundamentals, [and] threw all order and virtue out of the world." Locke's mistake was to make ideas and virtue and order "unnatural, and without foundation in our minds." Quoted in Norton, *From Moral Sense to Common Sense*, 72, 75.

61   Shaftesbury, "Sensus Communis," 81–82.

62   Ibid., 72.

63   Ibid., 96.

64   See Lawrence E. Klein, *Shaftesbury and the Culture of Politeness: Moral Discourse and Cultural Politics in Early Eighteenth-Century England* (Cambridge: Cambridge University Press, 1994).

65   Shaftesbury, "Sensus Communis," 89, 90.

66   Ibid., 102.

67   Ibid., 106.

68   Ibid., 106–7.

69   Ibid., 108–9.

70   This is the term Shaftesbury used most often in his "Inquiry" when discussing the moral sense, although he referred to the latter as well. See Shaftesbury, "An Inquiry Concerning Virtue, or Merit," 40–52.

71   Henry More, *A Collection of Several Philosophical Writings of Dr. Henry More, vol. 2* (London: James Fleisher, 1662), 70. More continues: "That which pinches us and vexes us so severely, is the sense that we have brought such an evil upon ourselves, when it was in our power to have avoided it." Shaftesbury was a follower of the Cambridge Platonists, publishing a collection of sermons by Benjamin Whichcote early in his literary career (1698).

72   Shaftesbury, "Sensus Communis," 132.

73   Ibid.

74   Ibid., 147. As he says in the *Inquiry Concerning Virtue*, virtuous behavior is the product both of a "Trial or Exercise of the Heart" and "sound and well-establish'd Reason" (30, 38).

75   As Stanley Grean puts it, "Every aspect of Shaftesbury's thought is dominated by his vision of the cosmos as a rational, coherent, effusive, and creative Whole." Grean, *Shaftesbury's Philosophy of Religion and Ethics*, 182. Shaftesbury's coherent cosmic vision is discussed in detail in "The Moralists, a Philosophical Rhapsody," published in Volume II of the *Characteristicks*.

76   Shaftesbury, "Sensus Communis," 61.

77   For documentation of the latter point see Jurgen Habermas, *The Structural Transformation of the Public Sphere: An Inquiry into a Category of Bourgeois Society* (Cambridge: MIT Press, 1989); J. A. W. Gunn, *Beyond Liberty and Property: The Process of Self-Recognition*

in Eighteenth-Century Political Thought (Kingston: McGill-Queen's University Press, 1983); Nicholas Rogers, *Whigs and Cities: Popular Politics in the Age of Walpole and Pitt* (Oxford: Oxford University Press, 1989); Wilson, *The Sense of the People*; and Geoff Eley, "Rethinking the Political: Social History and Political Culture in 18th and 19th Century Britain," *Archiv für Sozialgeschichte*, 21 (1981): 427–51.

78   Françoise Fénelon, *Démonstration de l'Existence de Dieu*, in Fénelon, *Œuvres II* (Paris: Gallimard, 1997), Troisième Preuve, 619–20.

79   See Joseph M. Levine, *The Battle of the Books: History and Literature in the Augustan Age* (Ithaca: Cornell University Press, 1991).

80   Shaftesbury, "Sensus Communis," 103–6.

81   John D. Schaeffer, *Sensus Communis: Vico, Rhetoric, and the Limits of Relativism* (Durham: Duke University Press, 1990), 2; and E. Vernon Arnold, *Roman Stoicism* (London: Routledge & Kegan Paul, 1958), 366–67.

82   *The New Science of Giambattista Vico* (1744), ed. and trans. Thomas Bergin and Max Frisch (Ithaca: Cornell University Press, 1970), p. 21, para. 142; p. 22, paras. 144–45.

83   Giambattista Vico, *The First New Science*, ed. and trans. Leon Pompa (Cambridge: Cambridge University Press, 2002), xxvi–xxx.

84   Ibid., 11, para. 10.

85   Ibid., 11, para. 11; 12, para. 13.

86   *The New Science of Giambattista Vico* (1744), 19–20, para. 129.

87   Vico, *The First New Science*, 30–31, para. 40.

88   For example, according to Amos Funkenstein Vico's common sense is "the definite mental configuration of each age, the harmonic principle of each period." Amos Funkenstein, *Theology and the Scientific Imagination: From the Middle Ages to the Seventeenth Century* (Princeton: Princeton University Press, 1986), 285–86. Leon Pompa similarly suggests that common sense for Vico means "the beliefs common to a society upon which its institutions and social order must stand. These vary at different points in the development of a nation." Vico, *The First New Science*, lviii. Cecilia Miller argues that for Vico *"sensus communis* was not static […] and has been adapted slightly in every culture." The question that remains, however, is *what* has been adapted? As Miller goes on to say, *sensus communis* is in fact "the shared attitudes of all men at all times," indicating the existence of a universal knowledge or wisdom that underlies particular manifestations of it, but which only becomes evident over time. See Cecilia Miller, *Giambattista Vico: Imagination and Historical Knowledge* (New York: St. Martin's Press, 1993), 30, 79–81.

89   *The New Science of Giambattista Vico* (1744), 25, para. 161.

90   John D. Schaeffer, *Sensus Communis: Vico, Rhetoric, and the Limits of Relativism* (Durham: Duke University Press, 1990), 84–85.

91   For Vico's relation to Cartesian thought see Miller, *Giambattista Vico*, Chapter 1.

92   This theme runs throughout Shaftesbury's *Sensus Communis*; for Vico, see the discussion in Schaeffer, *Sensus Communis*, Chapter 4.

Chapter 6 Common Sense and Moral Sense: Buffier, Hutcheson, and Butler

1   Kathleen S. Wilkins, *A Study of the Works of Claude Buffier* (Geneva: Institut et Musee Voltaire, 1969), 31.

2   For more see John N. Pappas, *Berthier's Journal de Trevoux and the Philosophes* (Geneva: Institut et Musee Voltaire, 1957); Cyril B. O'Keefe, S. J., *Contemporary Reactions to the Enlightenment (1728–1762): A Study of Three Critical Journals: The Jesuit*

Journal de Trévoux, *the Jansenist* Nouvelles ecclésiastiques, *and the secular* Journal des Savants (Geneva: Librairie Slatkine & Honoré Champion, 1974).

3   For the plagiarism charge See *First Truths, and the Origin or Our Opinions, Explained* [...] *Translated from the French of Pere Buffier. To Which Is Prefixed a Detection of the Plagiarism, Concealment, and Ingratitude of the Doctors Reid, Beattie, and Oswald* (London: J. Johnson, 1780). For comparison of the thought of Reid and Buffier, and consideration of the plagiarism charge, see Louise Marcil-Lacoste, *Claude Buffier and Thomas Reid: Two Common Sense Philosophers* (Kingston: McGill-Queen's University Press, 1982), 6–9, and Part IV.

4   Thomas Reid, *Essays on the Intellectual Powers of Man*, ed. Derek R. Brookes (University Park: Pennsylvania State University Press, 2002), 526.

5   Claude Buffier, *Traite des Premieres Veritez, et de la source de nos Jugemens, ou L'on examine le sentiment des Philosophes de ce temps, sur les premiéres notions des choses, par le P. Buffier, de la Compagnie de Jesus* (Paris: Chez la veuve Mauge, 1724), Book I, 3–5.

6   Ibid., vii–viii; Marcil-Lacoste, *Claude Buffier and Thomas Reid*, 14–15. In the *Elements of Metaphysics* Buffier emphasized the idea that philosophy, which should be accessible to everyone, involves critical thinking and the ability to abstract universal principles that lay behind the flux of appearances. See Claude Buffier, *Éléments de Métaphysique A la Portée De Tout Le Monde*, in *Oeuvres Philosophiques Du Père Buffier*, ed. Francisque Bouillier (Paris: Adolphe Delahays, 1843), esp. Conversations 1–3.

7   Buffier, *Traite des Premieres Veritez*, Book I, 10–11.

8   Ibid., Book I, 57–59; Marcil-Lacoste, *Claude Buffier and Thomas Reid*, 44–45.

9   Buffier, *Traite des Premieres Veritez*, Book I, 15.

10 Ibid., Book I, 23.

11 Ibid., Book I, 24–25.

12 Ibid., Book I, 26.

13 Ibid., Book I, 26–27, 38–39, 55; Buffier, *Éléments de* Métaphysique, 304–5.

14 As Kathleen Wilkins puts it, "Buffier's philosophical system [...] is essentially pragmatic. He professes to abandon sterile scholastic concepts and to build his theory of knowledge on data within the grasp of any reasonably intelligent man." Wilkins, *A Study of the Works of Claude Buffier*, 82.

15 Buffier, *Traite des Premieres Veritez*, Book I, 47–48, 247–48, 263–67; Buffier, *Éléments de Métaphysique*, 304.

16 One of Buffier's favorite first principles that gave clear support to the existence of God was the argument from design. According to Buffier, it is a first truth that chaos does not give rise to intelligence and design, for example, that a watch cannot be the product of chance but of an intelligent watchmaker. This was a common argument at the time and was understood to apply to the apparent intelligent design of the universe. See Buffier, *Traite des Premieres Veritez*, Book I, 26; Buffier, *Éléments de Métaphysique*, 304–5.

17 Buffier, *Traite des Premieres Veritez*, Book I, 27–32; Buffier, *Éléments de* Métaphysique, 283–84.

18 Buffier, *Traite des Premieres Veritez*, Book II, 258–59.

19 See the Epilogue for a literature review and discussion of the points of agreement between the philosophy of common sense and the findings of modern research.

20 For more, see the discussion in Wilkins, *A Study of the Works of Claude Buffier*, 89–95.

21 Buffier, *Traite des Premieres Veritez*, Book I, 117–18.

22 Ibid., Book I, 122–23.

23 Buffier, *Éléments de Métaphysique*, 305.

24 Ibid., 292, 299–300.

25 For example, if someone wants to deny the existence of first truths, ask him if he believes that you and he are separate persons. If he doesn't accept this as a first truth, then the only alternative is that you and he are the same person, and consequently there is no reason to speak further, since he will already know what you are going to say. Here and elsewhere, the bottom line is that rational discourse requires each person to accept certain fundamental truths at the outset, without demonstrative proof. Buffier, *Éléments de Métaphysique*, 304–5.

26 Buffier, *Traite des Premieres Veritez*, Book I, 55–57, 62.

27 Buffier, *Éléments de Métaphysique*, 294.

28 Ibid., 308.

29 Buffier, *Traite des Premieres Veritez*, Book I, 26; Buffier, *Éléments de Métaphysique*, 304.

30 Buffier, *Traite des Premieres Veritez*, Book I, 69.

31 Ibid., Book I, 67.

32 Marcil-Lacoste, *Claude Buffier and Thomas Reid*, 27. In emphasizing Buffier's anti-dogmatism and defense of paradoxes, Marcil-Lacoste argues that "Buffier actually makes it a methodological principle to deny that all men agree or should agree on most questions" (29). Here Lacoste overstates her case—Buffier first and foremost wants to assert, against skeptics, the existence of universally held first principles of common sense. His book is a "treatise on first truths," not a treatise on paradoxes and the plurality of opinion.

33 Buffier, *Traite des Premieres Veritez*, Book I, 68.

34 Ibid., Book I, 77–78.

35 Ibid., Book I, 69.

36 Ibid., Book I, 26. As we have seen, this is a view also held by Herbert of Cherbury.

37 Ibid., Book I, 88.

38 Ibid., Book II, 118–19.

39 For more on this point see the discussion in Marcil-Lacoste, *Claude Buffier and Thomas Reid*, 29–31.

40 Lacoste calls Buffier's thought a "mitigated and smiling form of skepticism." Ibid., 69. This may be true in regard to natural science, but it is unlikely Buffier would have accepted this designation in general, since in his view he was standing up for truth *against* skepticism.

41 Buffier, *Traite des Premieres Veritez*, Book I, 72–73; Book II, 105–8; Marcil-Lacoste, *Claude Buffier and Thomas Reid*, 60–65.

42 Buffier, *Traite des Premieres Veritez*, Book I, 94–99.

43 Ibid., Book II, 98.

44 Ibid., Book II, 106; 103.

45 Ibid., Book II, 108.

46 Ibid., Book I, 114.

47 Ibid., Book I, 103–4.

48 Wilkins, *A Study of the Works of Claude Buffier*, 95.

49 Buffier's philosophical ideas were cited most extensively in the article "Sens," where segments of the *Traité* are quoted verbatim, at length and without attribution. His general doctrine of *sens comun* is not reproduced in the article, however, and this may be one reason why nineteenth-century French philosophers knew Reid better than Buffier. For more on Buffier's influence on the French Enlightenment and his contribution to the *Encyclopédie*, see the discussion in Wilkins, *A Study of the Works of Claude Buffier*, 97–115, and Appendix C.

50   Ibid., 90–101, 115. As Buffier put it,

> Mr. Locke is the first of the moderns who has sought to investigate the
> operations of the human mind directly from nature, without being biased by
> opinions that are the product of systems rather than real evidence. His philos-
> ophy, when compared to that of Des Cartes and Malebranche appears to be as
> superior to either, as history is to romance.

Traite des Premieres Veritez, Book II, 253.

51   For discussion of Buffier's affinity for Descartes, which was fairly unusual for Jesuits at
the time, as well as his disagreements with the great philosopher, see Wilkins, *A Study of
the Works of Claude Buffier*, 68–70, 85–89, 94–96. See also Marcil-Lacoste, *Claude Buffier
and Thomas Reid*, 44–69.

52   Buffier did have an impact on Catholic thought, as, for example, on the Spanish
philosopher Jaime Balmes, who flourished in the first half of the nineteenth century. See Kelly James Clark, "Spanish Common Sense Philosophy: Jaime Balmes'
Critique of Cartesian Foundationalism," *History of Philosophy Quarterly*, 7, no. 2 (April
1990): 207–26.

53   The editor of the 1822 edition of Buffier's *Traité* notes that Buffier was not widely
referenced in current debates about common sense and surmised that this was due to
the inaccessibility of the original text. See Claude Buffier, *La Doctrine du sens commun,
ou Traité des premières vérités et de la source de nos jugemens, suivi d'une exposition des preuves les
plus sensibles de la véritable réligion* (Avignon: Séguin-Aîné, 1822), 4–5. Given that he was
in his time a well-known author and that the *Éléments de Métaphysique* was an accessible
account of his ideas, this explanation seems unlikely. It is more probable that his philosophical works had simply been forgotten by the time the soil was fertile for them in
the early nineteenth century.

54   Particularly relevant to the present discussion is Daniel Gordon, *Citizens without
Sovereignty: Equality and Sociability in French Thought, 1670–1789* (Princeton: Princeton
University Press, 1994). See also Keith Michael Baker, "Defining the Public Sphere
in Eighteenth-Century France: Variations on a Theme by Habermas," in *Habermas
and the Public Sphere*, ed. Craig Calhoun (Cambridge: Cambridge University Press,
1992), 181–211; Dena Goodman, *The Republic of Letters: A Cultural History of the French
Enlightenment* (Ithaca, NY: Cornell University Press, 1994); Mona Ozouf, "'Public
Opinion' at the End of the Old Regime," *Journal of Modern History*, 60, suppl.
(September 1988): S1–S21; Roger Chartier, *The Cultural Origins of the French Revolution*
(Durham: Duke University Press, 1991). See also Jürgen Habermas, *The Structural
Transformation of the Public Sphere: An Inquiry into a Category of Bourgeois Society* (Cambridge,
MA: MIT Press, 1962, 1989).

55   Quoted in Daniel Gordon, *Citizens without Sovereignty: Equality and Sociability in French
Thought, 1670–1789* (Princeton: Princeton University Press, 1994), 79. For detailed
discussion of Buffier's ethical theory see Wilkins, *A Study of the Works of Claude Buffier*,
117–41.

56   E.g. Buffier, *Traite des Premieres Veritez*, Book I, 56, 67, 75.

57   See the discussion of Buffier in Gordon, *Citizens without Sovereignty*, 79–81.

58   See Thomas Lockwood, "The Life and Death of *Common Sense*," in *Telling People What
to Think: Early Eighteenth-Century Periodicals from* The Review *to* The Rambler, ed. J. A.
Downie and Thomas N. Corns (London: Frank Cass, 1993), 78–93.

59   Henry Fielding, *Pasquin. A Dramatick Satire on the Times: Being the Rehearsal of Two Plays, viz. A Comedy call'd, The Election; and a Tragedy call'd, The Life and Death of Common-Sense. As It Is Acted at the Theatre in the Hay-Market* (London: J. Watts, 1736), 49, 51, 56, 64.

60   *Common Sense: Or, The Englishman's Journal* 66, May 6, 1738. Thomas Lockwood sees the journal, at least in its early years, as a conscious revival of *The Spectator*. Lockwood, "The Life and Death of *Common Sense*," 90.

61   *Common Sense: Or, The Englishman's Journal* 1, February 5, 1737.

62   Ibid., 2, February 12, 1737; ibid., 16, May 21, 1737; ibid., 24, July 16, 1737; ibid., 42, November 19, 1737.

63   Ibid., 66, May 6, 1738; see also ibid., 2, February 12, 1737; ibid., 24, July 16, 1737.

64   Ibid., 13, April 30, 1737.

65   Ibid., 4, February 26, 1737; ibid., 71, June 10, 1738.

66   Ibid., 66, May 6, 1738.

67   Sophia Rosenfeld, *Common Sense: A Political History* (Cambridge: Harvard University Press, 2011), 29–30.

68   Thomas Reid, *Essays on the Active Powers of Man* (London: G.G.J. & J. Robinson, 1788), 236–40; MS 3061/10, Special Collections, Aberdeen University Library.

69   The Scottish Enlightenment is discussed further in Chapters 7 and 8.

70   For Mandeville's impact on Butler see E. J. Hundert, *The Enlightenment's Fable: Bernard Mandeville and the Discovery of Society* (Cambridge: Cambridge University Press, 1994), 126–32.

71   Hundert's *The Enlightenment's Fable* remains the definitive study of Mandeville in his European context.

72   See "An Enquiry into the Origin of Moral Virtue," in Bernard Mandeville, *The Fable of the Bees: Or, Private Vices, Publick Benefits Vol. 1*, ed. F. B. Kay (Indianapolis: Liberty Classics, 1988). Quote from p. 48.

73   Bernard Mandeville, "A Search into the Nature of Society," in *The Fable of the Bees: Or, Private Vices, Publick Benefits Vol. 1*, ed. F. B. Kay (Indianapolis: Liberty Classics, 1988), 324.

74   Ibid., 332.

75   Ibid., 333.

76   Ibid., 324.

77   Ibid., 326–31.

78   Ibid., 336–41; quote from 340.

79   Ibid., 341.

80   Ibid., 341–42.

81   Hundert, *The Enlightenment's Fable*, 13.

82   Francis Hutcheson, *An Essay on the Nature and Conduct of the Passions and Affections, with Illustrations on the Moral Sense* (Indianapolis: Liberty Fund, 2002), *passim*; quotes from 135–36. See also Francis Hutcheson, *A System of Moral Philosophy* (London: A. Millar, 1755), esp. 24–25, 56–79; Francis Hutcheson, *Inaugural Lecture on the Social Nature of Man*, in Francis Hutcheson, *On Human Nature*, ed. Thomas Mautner (Cambridge: Cambridge University Press, 1993), 138–43.

83   Hutcheson, *An Essay on the Nature and Conduct of the Passions and Affections*, 136, 173–75; Hutcheson, *A System of Moral Philosophy*, 1, 41.

84   Hutcheson, *An Essay on the Nature and Conduct of the Passions and Affections*, 178; See also Daniel Carey, "Method, Moral Sense, and the Problem of Diversity: Francis Hutcheson and the Scottish Enlightenment," *British Journal for the History of Philosophy*, 5, no. 2

NOTES
191

(1997): 275–96; David Fate Norton, *From Moral Sense to Common Sense: An Essay on the Development of Scottish Common Sense Philosophy, 1700–1765* (PhD Dissertation, University of California at San Diego, 1966), 117–19.

85 Hutcheson, *An Essay on the Nature and Conduct of the Passions and Affections*, 9; see also 106, 126.

86 Hutcheson was very open about this. In his *Inaugural lecture on the social nature of man*, given to the University of Glasgow in 1730, Hutcheson framed his project as an inquiry "into what kinds of things can rightly be called natural to man as far as morality is concerned, and then, to what extent society, be it civil society or society under no human authority, can be counted among things natural" (127).

87 See, for example, Hutcheson's reply to Pufendorf in his *Inaugural Lecture on the Social Nature of Man*, 134–36. As indicated in the previous note, Hutcheson opened this lecture by stressing that his inquiry was a purely naturalistic one.

88 Samuel Pufendorf, *On the Duty of Man and Citizen According to Natural Law*, trans. Michael Silverthorne (Cambridge: Cambridge University Press, 1991), 35. See also the discussion in James Moore, "The Two Systems of Francis Hutcheson: On the Origins of the Scottish Enlightenment," in *Studies in the Philosophy of the Scottish Enlightenment*, ed. M. A. Stewart (Oxford: Clarendon Press, 1990), 37–59; Richard Tuck, *Natural Rights Theories: Their Origin and Development* (Cambridge: Cambridge University Press, 1979); Knud Haakonssen, *Natural Law and Moral Philosophy: From Grotius to the Scottish Enlightenment* (Cambridge: Cambridge University Press, 1996); and Alexander Passerin d'Entrèves, *Natural Law: An Introduction to Legal Philosophy* (New Brunswick: Transaction Publishers, 1994), esp. Chap. 4.

89 Francis Hutcheson, *Reflections upon Laughter, and Remarks upon the Fable of the Bees* (Glasgow: MDCCL, [1725] 1750), 7.

90 For a good example of Hutcheson's scientism see *A System of Moral Philosophy*, Book I, Chap. 1.

91 Hutcheson, *Inaugural Lecture on the Social Nature of Man*, 144.

92 For discussion of the similarities between Hutcheson and Reid on this score see Norton, *From Moral Sense to Common Sense*, 122–24.

93 Butler mainly elaborates on this moral sense in his appended essay "Of the Nature of Virtue." There he wrote of a "faculty of discernment" that naturally and unavoidably approves of some actions as being virtuous or good and disapproves of others as being vicious or bad. This faculty, "whether called conscience, moral reason, moral sense, or divine reason," is manifested throughout the world in common language and behavior. Joseph Butler, *The Analogy of Religion, Natural and Revealed, to the Constitution and Course of Nature* (New York: New and Co., 1851), 264–69.

94 Ibid., 40.
95 Ibid., 42.
96 Ibid., 255.
97 Ibid., 251.
98 Ibid., 256.
99 Ibid., 119.
100 Ibid., 135–37.
101 Ibid., 261.
102 Ibid., 262–63.
103 Ibid., 263.
104 Ibid., 262.
105 Ibid., 137.

Chapter 7 Common Sense and the Science of Man in Enlightenment Scotland: Turnbull and Kames

1   Reid's impact and influence in nineteenth-century Britain, France, Germany, and America is discussed at length in the Epilogue.

2   Donald J. Withrington, "What Was Distinctive about the Scottish Enlightenment?," in *Aberdeen and the Enlightenment*, ed. Jennifer J. Carter and Joan H. Pittock (Aberdeen: Aberdeen University Press, 1987), 15.

3   Alexander Broadie, *The Scottish Enlightenment: The Historical Age of a Historical Nation* (Edinburgh: Birlinn, 2001), 25. See also Alexander Broadie, *The Tradition of Scottish Philosophy: A New Perspective on the Enlightenment* (Savage, MD: Barnes and Noble Books, 1990); Alexander Broadie, ed., *The Cambridge Companion to the Scottish Enlightenment* (Cambridge: Cambridge University Press, 2003); and David Allan, *Virtue, Learning and the Scottish Enlightenment: Ideas of Scholarship in Early Modern History* (Edinburgh: Edinburgh University Press, 1993).

4   Roger Emerson, "The Contexts of the Scottish Enlightenment," in *The Cambridge Companion to the Scottish Enlightenment*, 17–18.

5   Anand C. Chitnis, "The Eighteenth-Century Scottish Intellectual Inquiry: Context and Continuities versus Civic Virtue," in *Aberdeen and the Enlightenment*, 79; see also M. A. Stewart, "Religion and Rational Theology," in *The Cambridge Companion to the Scottish Enlightenment*, 31–59.

6   Nicholas Phillipson, "The Scottish Enlightenment," in *The Enlightenment in National Context*, ed. Roy Porter and Mikulas Teich (Cambridge: Cambridge University Press, 1981), 20.

7   Anand C. Chitnis, "The Eighteenth-Century Scottish Intellectual Inquiry: Context and Continuities versus Civic Virtue," in *Aberdeen and the Enlightenment*, 87. Broadie shares this view, suggesting that "science is crucially important to the Scottish Enlightenment, perhaps more important than were moral philosophy, historiography and political economy." The scientific work of figures like William Cullen, James Watt, and Joseph Black was complemented by scientific ways of thinking that "were at work across the whole range of intellectual disciplines." Alexander Broadie, "Introduction," in *The Cambridge Companion to the Scottish Enlightenment*, 4.

8   Paul Wood, "Science in the Scottish Enlightenment," in *The Cambridge Companion to the Scottish Enlightenment*, 95. See also Roger Emerson, "Science and Moral Philosophy in the Scottish Enlightenment," in *Studies in the Philosophy of the Scottish Enlightenment*, ed. M. A. Stewart (Oxford: Clarendon Press, 1990), 11–36; Charles Withers and Paul Wood, eds., *Science and Medicine in the Scottish Enlightenment* (East Linton, Scotland: Tuckwell Press, 2002); and Paul Wood, *The Aberdeen Enlightenment: The Arts Curriculum in the Eighteenth Century* (Aberdeen: Aberdeen University Press), 1993.

9   George Turnbull, *Observations upon Liberal Education, in All Its Branches* (Indianapolis: Liberty Fund, 2003), 127. See also 199.

10   George Turnbull, *The Principles of Moral and Christian Philosophy, vol. 1* (Indianapolis: Liberty Fund, 2005), 12–17.

11   See David Fate Norton, "George Turnbull and the Furniture of the Mind," *Journal of the History of Ideas*, 36, no. 4 (1975): 701–16. See also David Fate Norton, *From Moral Sense to Common Sense: An Essay on the Development of Scottish Common Sense Philosophy, 1700–1765* (San Diego: University of California, PhD Dissertation). Turnbull neatly summarizes our "furniture for acquiring knowledge" in his *Observation upon Liberal*

Education thus: "Call to mind our senses, our memory, our reason, our imagination, and together with them our sense of harmony and beauty, our delight in comparing effects, and in tracing them to general laws—all of which powers are capable of [...] high improvement." Turnbull, *Observations upon Liberal Education*, 118. For discussion of Henry More's use of the term, see Chapter 4. Turnbull cites More's *Divine Dialogues* a number of times in *The Principles*, where he calls him "a very good author." See Turnbull, *The Principles of Moral and Christian Philosophy, vol. 1*, 306, 320, 371. Quote from 306.

12   Limitations of space prevent me from examining the whole of Turnbull's corpus, which includes religious writings (e.g., *A Philosophical Enquiry concerning the Connexion betwixt the Doctrines and Miracles of Jesus Christ*, 1731, 1732, 1739), educational treatises (e.g., *Observations upon Liberal Education*, 1742), and commentary on moral and civil law (*A Discourse upon the Nature and Origine of Moral and Civil Laws*, 1740). Suffice it to say that Turnbull's philosophical perspective discussed here was evident in all his works.

13   Reid for his part would go on to develop a philosophy of science—drawing on Newton's methodology—that demarcated physics as an autonomous science from metaphysics, theology, and moral philosophy, a distinction that Turnbull did not make. See Robert Callergård, *An Essay on Thomas Reid's Philosophy of Science* (Stockholm: Stockholm University, 2006). Callergård demonstrates that Reid's science of the mind "resembles botany and chemistry more than classical mechanics. Rather than reduction of phenomena to general laws, prediction, and explanation, the science of the mind is concerned with identification, classification, and analysis of components into simples. Reid's favorite simile is that of 'anatomy of the mind,' and anatomy bears little resemblance [to] classical mechanics." Ibid., 60.

14   George Turnbull, *Education for Life: Correspondence and Writings on Religion and Practical Philosophy*, ed. M. A. Stewart and Paul Wood, trans. Michael Silverthorne (Indianapolis: Liberty Fund, 2014), 49, 51, 52. As we will see, Reid developed a similar idea of causality as natural law; and he developed the idea, already present in Turnbull's thinking (discussed further below), that we have knowledge of such causes via a principle of common sense.

15   Turnbull, *Education for Life*, 54.

16   It should be noted at the outset that Turnbull himself used the term "common sense" infrequently and without the philosophical precision employed by his pupil Thomas Reid. Here I will highlight his usage of the term in his *Principles of Moral and Christian Philosophy*. For a broader survey of his use of the term see Norton, *From Moral Sense to Common Sense*, Chapter VI.

17   Turnbull, *Education for Life*, 70–73.

18   For overviews of Turnbull's life and work as an educator see M. A. Stewart, "George Turnbull and Educational Reform," in *Aberdeen and the Enlightenment*, 95–103; Paul Wood and M. A. Stewart, "Introduction," in *Education for Life*, ix–xxvi; Alexander Broadie, "Introduction," in *The Principles of Moral and Christian Philosophy, vol. 1*, ix–xvii.

19   Turnbull, *The Principles of Moral and Christian Philosophy, vol. 1*, 5, 8.

20   Turnbull, *The Principles of Moral and Christian Philosophy, vol. 1*, 8–9. As Turnbull states later in the book,

All indeed I have any right to pretend to, is to have attempted to dispose very ancient observations upon mankind and moral providence, into the order that natural philosophers, after Sir Isaac Newton, follow, in accounting for material phenomena [...] It is in knowledge of the natural world that we surpass the ancients [...who] accounted for moral effects, by reducing them to powers and their laws, or manners of operation.

Turnbull, *The Principles of Moral and Christian Philosophy, vol. 1*, 439.

21   For more on the natural law tradition as it relates to Turnbull, see Knud Haakonssen, *Natural Law and Moral Philosophy: From Grotius to the Scottish Enlightenment* (Cambridge: Cambridge University Press, 1996), esp. 85–90; Knud Haakonssen, "Natural Law and Moral Realism: The Scottish Synthesis," in *Studies in the Philosophy of the Scottish Enlightenment*, 61–85; and K. A. B. Mackinnon, "George Turnbull's Common Sense Jurisprudence," in *Aberdeen and the Enlightenment*, 104–10.

22   Turnbull often makes reference to Cicero's writings when discussing the natural laws of human and physical nature, and our ability to know them via our reason and senses. For example, Turnbull quotes long passages from Cicero's *de natura Deorum* to support the statement that "What is greater, or more elevating, than the contemplation of nature, when we are able to take in large views of it, and comprehend its laws? How agreeably do ancient philosophers expatiate on this topic!" Turnbull, *The Principles of Moral and Christian Philosophy, vol. 1*, 104–5; see also, for example, 54–55, 84, 90, 143, 182–83, 208–9, 230–34. Throughout the text Turnbull echoes the view, expressed by Cicero in *De Re Publica* and other writings, that

True law is right reason in agreement with nature; it is of universal application, unchanging and everlasting; it summons to duty by its commands, and averts from wrongdoing by its prohibitions [...] And there will not be different laws at Rome and at Athens, or different laws now and in the future, but one eternal and unchangeable law will be valid for all nations and all times, and there will be one master and ruler, that is, God, over us all, for he is the author of this law, its promulgator, and its enforcing judge. Whoever is disobedient is fleeing from himself and denying his human nature.

De Re Publica, II, XXII, in *Cicero: De Re Publica, De Legibus, with an English Translation by Clinton Walker Keyes* (Cambridge: Harvard University Press and William Heinemann, 1959), 211. See also the discussion in Jed W. Atkins, *Cicero on Politics and the Limits of Reason* (Cambridge: Cambridge University Press, 2013), 165–76.

23   Turnbull, *The Principles of Moral and Christian Philosophy, vol. 1*, 438.

24   Ibid., 59.

25   Ibid., 78.

26   Turnbull, *The Principles of Moral and Christian Philosophy, vol. 2* (Indianapolis: Liberty Fund, 2005), 567.

27   Ibid., 568–71.

28   Turnbull, *The Principles of Moral and Christian Philosophy, vol. 1*, 79–80.

29   Ibid., 82.

30   Ibid., 119–21. Turnbull thus obviously follows Locke's theory of the association of ideas—which Reid would reject—as a way to understand how we are able to acquire higher-order knowledge, while developing the prerational bases for knowledge that remained as lacunae in Locke's theory. As we have seen such lacunae were already being attacked by earlier thinkers in a similar fashion (see Chapter 5).

31   Ibid., 124.

32   Ibid., 224.

33   Ibid., 125ff.

34   As indicated above, Reid's philosophy of science was more nuanced. While Reid carried forward the general project of applying the methodology of Newton and Bacon to areas of inquiry beyond physics—including study of the human mind—he differentiated physics from other sciences and strongly rejected the idea that morals could be subject to quantification. See Callergård, *An Essay on Thomas Reid's Philosophy of Science*.

35   Turnbull, *The Principles of Moral and Christian Philosophy, vol. 1*, 149.

36   Ibid., 152, 168.

37   Ibid., 169, 171.

38   Turnbull, *The Principles of Moral and Christian Philosophy, vol. 2*, 579.

39   Turnbull, *The Principles of Moral and Christian Philosophy, vol. 1*, 176–77.

40   George Turnbull, *A Treatise on Ancient Painting, Containing Observations on the Rise, Progress, and Decline of that Art amongst the Greeks and Romans* (London: A. Millar, 1740), xii. According to Terrence Moore, in embracing subjects "that Locke either criticized or largely ignored, such as poetry, painting, and the natural sciences […] Turnbull was one of the first educational theorists to open up liberal education to the study of the natural world as it had been explained by Newton. He stood virtually alone in his enthusiasm for art." Terrence O. Moore, "Introduction," in *Observations upon Liberal Education*, xv.

41   For development of this latter idea in relation to Turnbull see David Allan, *Virtue, Learning and the Scottish Enlightenment: Ideas of Scholarship in Early Modern History* (Edinburgh: Edinburgh University Press, 1993), 174–200.

42   For more on Kames's life and letters see William C. Lehmann, *Henry Home, Lord Kames, and the Scottish Enlightenment: A Study in National Character and in the History of Ideas* (The Hague: Martinus Nijhoff, 1971); and Arthur E. McGuinness, *Henry Home, Lord Kames* (New York: Twayne, 1970).

43   As he put it in *Elements of Criticism*, when pursuing the science of human nature, the way to avoid "utopian systems" and "the fatigue of reasoning" is "to take a survey of human nature, and to set before the eye, plainly and candidly, facts as they really exist." From facts and experiments one then "ascend[s] gradually to principles." Henry Home and Lord Kames, *Elements of Criticism, vol. 1*, ed. Peter Jones (Indianapolis: Liberty Fund, 2005), 33, 18.

44   Henry Home and Lord Kames, *Essays on the Principles of Morality and Natural Religion* (Edinburgh: Kincaid & Donaldson, 1751), advertisement.

45   Ibid., 27.

46   Ibid., 66–68.

47   Ibid., 154–55. Elsewhere, Kames states:

>   As in the natural world, the Almighty has adapted our senses, not to the dis-
>   covery of the intimate nature and essences of things, but to the uses and
>   conveniences of life […] exhibit[ing] natural objects to us, not in their real, but
>   in a sort of artificial view […] so he has exhibited the intellectual world to us,
>   in a like artificial view, clothed with certain colours and distinctions, imaginary,
>   but useful.

Ibid., 190. Here the incipient pragmatism of common-sense philosophy is evident. For discussion of the pragmatic elements of Kames's philosophy see Lehmann, *Henry Home, Lord Kames, and the Scottish Enlightenment*, 167–71.

48   Kames, *Essays on the Principles of Morality and Natural Religion*, 156–61, 187.

49   Ibid., 164–67.

50   Ibid., 188–90.

51   See Norton, "From Moral Sense to Common Sense," 236–76. See especially pp. 271–75 for detailed comparison of the thought of Reid and Kames.

52   Kames, *Essays on the Principles of Morality and Natural Religion*, 227–29.

53   Thomas Reid, *An Inquiry into the Human Mind, on the Principles of Common Sense*, ed. Derek Brookes (University Park: Pennsylvania State University Press, 1997), 193–94.

54   Kames, *Essays on the Principles of Morality and Natural Religion*, 234–35.

55   Ibid., 39–40.

56   Reid, *An Inquiry into the Human Mind*, 169. Reid repeatedly argues that a fundamental mistake made by philosophers was to give undue privilege to discursive reason over other faculties in our sense-making of the world, leading to skepticism and absurd conclusions that only disgrace philosophy. This mistake is doubly egregious, in Reid's view, because rational, scientific thought has its starting point in the first principles of common sense: "Philosophy[…] has no other root but the Principles of Common Sense; it grows out of them, and draws its nourishment from them: severed from this root, its honours wither, its sap is dried up, it dies and rots" (19).

57   McGuinness, *Henry Home, Lord Kames*, 36.

58   Kames, *Essays on the Principles of Morality and Natural Religion*, 242.

59   Ibid., 242–43.

60   Ibid., 252.

61   Ibid., 263–65.

62   Ibid., 275–79.

63   Ibid., 282.

64   Henry Home, *Introduction to the Art of Thinking* (Edinburgh: William Creech, 1789), 97–99.

65   Kames, *Essays on the Principles of Morality and Natural Religion*, 282–89.

66   Ibid., 303.

67   Reid, *An Inquiry into the Human Mind*, 198–200.

68   Kames, *Essays on the Principles of Morality and Natural Religion*, 305–6.

69   As Kames put it in *The Art of Thinking*, in a section titled "Our Opinions are swayed more by Feeling than by Argument": "That reason which is favourable to our desires, appears always the best." Home, *Introduction to the Art of Thinking*, 6.

70   Kames, *Essays on the Principles of Morality and Natural Religion*, 289.

71   Ibid., 284.

72   Lehmann, *Henry Home, Lord Kames, and the Scottish Enlightenment*, 132.

73   Home, *Introduction to the Art of Thinking*, 40–41.

74   See Lehmann, *Henry Home, Lord Kames, and the Scottish Enlightenment*, 137–38. See also Home, *Introduction to the Art of Thinking*, 7, 51. Turnbull for his part indicated that although we are all endowed with the same basic mental furniture, social divisions are for the general good, and "there is no such thing as a perfect democracy." See Turnbull, *Observations upon Liberal Education*, 119–21.

75   Home, *Introduction to the Art of Thinking*, 9.

76   Ibid., 7.

77   Lehmann, *Henry Home, Lord Kames, and the Scottish Enlightenment*, 138.

78   Ibid.

Chapter 8 Common Sense, Science, and the Public Sphere: The Philosophy of Thomas Reid

1   For the impact of the common-sense school in America see Terence Martin, *The Instructed Vision: Scottish Common-Sense Philosophy and the Origins of American Fiction* (Bloomington: Indiana University Press, 1961). For Reid's impact in Britain and on the Continent see (among others) James McCosh, *The Scottish Philosophy* (London: Macmillan, 1875); Manfred Kuehn, *Scottish Common Sense in Germany, 1768–1800: A Contribution to the History of Critical Philosophy* (Kingston: McGill-Queen's University Press, 1987); James W. Manns, *Reid and His French Disciples: Aesthetics and Metaphysics* (Leiden: Brill, 1994). The Epilogue presents a synthetic overview of Reid's influence and relevance today.

2   For in-depth treatment of Reid's philosophy of science, see Robert Callergård, *An Essay on Thomas Reid's Philosophy of Science* (Stockholm: Stockholm University, 2006).

3   A Campbell Fraser, *Thomas Reid* (Edinburgh: Oliphant Anderson & Ferrer, 1898), remains a useful source for basic information about Reid.

4   Thomas Reid, *Essays on the Intellectual Powers of Man*, ed. Derek R. Brookes (University Park, PA: Pennsylvania State University Press, 2002), 423–25. Hereafter cited as *Intellectual Powers*.

5   *Intellectual Powers*, 425–26.

6   Ibid., 426.

7   Ibid., 426–27.

8   Ibid., 427.

9   Reid most likely became aware of Buffier's work some time after he had developed his own philosophy of common sense. Reid mentions, for example, in "A Brief Account of Aristotle's Logic," first published in 1774, that "I have lately met with a very sensible and judicious treatise, written by Father Buffier about fifty years ago, concerning first principles […]." Given that Buffier's *Treatise on First Truths and the Source of Our Judgments* was published in 1724, and that Reid was, by all accounts and evidence, a modest and humble person who would not lie about such things, it seems very likely that Reid was not aware of Buffier's writings until after he published his *Inquiry* in 1764. See *The Works of Thomas Reid, with an Account of His Life and Writings by Dugald Stewart, vol. 1* (Charlestown: Samuel Etheridge Junior, 1813), 163–64. See also the discussion in Louise Marcil-Lacoste, *Claude Buffier and Thomas Reid: Two Common Sense Philosophers* (Kingston: McGill-Queen's University Press, 1982), 6–9, and Part IV.

10 *Intellectual Powers*, 427. Reid was skeptical of Berkeley's claim that his philosophy was in accord with common sense: "It is pleasant to observe the fruitless pains which Bishop Berkeley takes to shew, that his system of the non-existence of a material world did not contradict the sentiments of the vulgar, but those only of the philosophers." Ibid., 464.

11 Ibid., 428–29.

12 Ibid., 430–31.

13 Ibid., 431–32.

14 Ibid., 13.

15 See, for example, the extended discussion/critique of the theory of ideas in *Intellectual Powers*, 125–87.

16 Ibid., 432–33.

17 Ibid., 433–34.

18   Ibid., 452–53.
19   Ibid., 471–90.
20   Ibid., 490–97.
21   "Thomas Reid's Curâ Primâ on Common Sense" edited by David Fate Norton, In Louise Marcil-Lacoste, Claude Buffier and Thomas Reid: Two Common Sense Philosophers (Kingstons: McGill-Queen's University Press, 1982), 187–208. Norton dates the fragment to late 1768 or early 1769.
22   See the Epilogue.
23   There is a large and growing body of work that analyzes the thought of Reid in the detail it deserves, including, in addition to journal articles and the journal Reid Studies, the following books: Stephen F. Barker and Tom L. Beauchamp, eds., Thomas Reid: Critical Interpretations (Philadelphia: Philosophical Monographs, 1976); Rebecca Copenhaver and Todd Buras, eds., Thomas Reid on Mind, Knowledge, and Value (New York: Oxford University Press, 2015); Terence Cuneo and René van Woudenberg, eds., The Cambridge Companion to Thomas Reid (Cambridge: Cambridge University Press, 2004); Roger D. Gallie, Thomas Reid and the Way of Ideas (Dordrecht, Netherlands: Kluwer, 1989); Joseph Houston, ed., Thomas Reid: Context, Influence, Significance (Edinburgh: Dunedin Academic Press, 2004); Keith Lehrer, Thomas Reid (New York: Routledge, 1989); Ryan Nichols, Reid's Theory of Perception (Oxford: Oxford University Press, 2007); Nicholas Wolterstorff, Thomas Reid and the Story of Epistemology (Cambridge: Cambridge University Press, 2001). The aim of the present chapter is to provide a contextual reading of some of Reid's foundational ideas and to locate his thought within the history of ideas, in particular the historical interplay between common sense and science, rather than providing a comprehensive overview of his thought.
24   David Hume, A Treatise of Human Nature, ed. L. A. Selby-Bigge (Oxford: Clarendon Press, 1978), 10. Because of the wide range and number of cited pages, subsequent page citations to this book will be made in the body of the text.
25   For recent discussions of Hume and the "problem of induction" in the early modern period, see Peter Dear, Discipline and Experience: The Mathematical Way in the Scientific Revolution (Chicago: University of Chicago Press, 1995), Introduction and Chapter 1; J. R. Milton, "Induction before Hume," British Journal for the Philosophy of Science, 38 (1987): 49–74; Mary Poovey, A History of the Modern Fact: Problems of Knowledge in the Sciences of Wealth and Society (Chicago: University of Chicago Press, 1998), 173–75. This notion is discussed further below.
26   Thus although Donald Livingston argues that Hume entertained a dialectical "philosophy of common life" that incorporated the perceptions of everyday life as the indispensable ground for skeptical inquiry, he recognizes that "The idea of common life and its special authority is not [according to Hume] an idea available to the vulgar, who are sunk in its unreflective order." Thus Hume does not "claim to find in the order of common life a new foundation of true propositions on which philosophy could be built, as Thomas Reid and the Scottish school of common sense claimed to have found." Donald W. Livingston, Hume's Philosophy of Common Life (Chicago: University of Chicago Press, 1984), 30.
27   Harvey Chisick, "David Hume and the Common People," in The "Science of Man" in the Scottish Enlightenment: Hume, Reid and Their Contemporaries, ed. Peter Jones (Edinburgh: Edinburgh University Press, 1989), 5–32. See also Livingston, Hume's Philosophy of Common Life, 30–31. In his essay "Of Commerce," Hume argued that

commoners, in contrast to philosophers, base their thinking on "shallow" principles and are unable to distinguish particular from general circumstances. David Hume, *Essays: Moral, Political, and Literary*, ed. Eugene F. Miller (Indianapolis: Liberty Classics, 1987), 254. Hume also felt that Reid had, in the *Inquiry*, imagined "the Vulgar to be Philosophers and Corpuscularians from their Infancy," at least in their ability to understand the difference between sensible qualities like heat and their physical causes. See P. B. Wood, "David Hume on Thomas Reid's *An Inquiry into the Human Mind, On the Principles of Common Sense*: A New Letter to Hugh Blair from July 1762," *Mind*, 95, no. 380 (October 1986): 416.

28   R. L. Woolhouse, *The Empiricists* (Oxford: Oxford University Press, 1988), 6–7. See also Richard Campbell, *Truth and Historicity* (Oxford: Clarendon Press, 1992), 147–8; Ilpo Halonen and Jaako Hintikka, "Aristotelian Explanations," *Studies in the History and Philosophy of Science*, 31, no. 1 (2000): esp. 125–28; Max Hocutt, "Aristotle's Four Becauses," *Philosophy*, 49/190 (1974): 385–99.

29   Aristotle, *Metaphysics*, in *The Basic Works of Aristotle*, ed. Richard McKeon (New York: Random House, 1941), Book I, Chap. 1, 980–82, pp. 690–92; Book II, Chap. 1, 993, p. 712.

30   Aristotle, *Nicomachean Ethics*, in *The Basic Works of Aristotle*, Book IV, Chap. 3, 1024–25, p. 1139. Discussed at length in Chapter 1.

31   See Michael Frede, "The Original Notion of Cause," in *Doubt and Dogmatism: Studies in Hellenistic Epistemology*, ed. Macolm Shofield, Myles Burnyeat, and Jonathan Barnes (Oxford: Clarendon Press, 1980), 217–49; Richard Sorabji, "Causation, Laws, and Necessity," in *Doubt and Dogmatism: Studies in Hellenistic Epistemology*, 250–82. Both of these authors deal especially with Stoic departures from Aristotelian understandings of causality, in which the sense of "cause" as active agent that produces a change in another entity becomes paramount. Frede argues that this Stoic narrowing of the meaning of cause contributed to the modern understanding of "cause" as *efficient* cause (rather than any of the other Aristotelian becauses).

32   Woolhouse, *The Empiricists*, 19. As the Father of Salomon's House put it in *New Atlantis*, "The end of our Foundation is the Knowledge of Causes, and secret motions of things; and the enlarging of the bounds of Human Empire, to the effecting of all things possible." Francis Bacon, *New Atlantis*, in *Francis Bacon: A Critical Edition of the Major Works*, ed. Brian Vickers (Oxford: Oxford University Press, 1996), 480. This topic is discussed at length in Chapter 1.

33   Stephen Buckle argues that Hume entertained a "romantic" notion of scientific truth (a notion that fed his skepticism), believing that "the systematic knowledge of necessary connections—philosophical knowledge as *scientia*—is the standard at which philosophy aims, and by which philosophies are judged." Stephen Buckle, "British Sceptical Realism: A Fresh Look at the British Tradition," *European Journal of Philosophy*, 7, no. 1 (1999): 16.

34   Many forms of skepticism derived in part from the vagaries of sense experience; and as Milton demonstrates, the fallibility of inductive (sensory) inferences had been foregrounded early on by the Stoics, and doubts about the reliability of inductive inferences had been present from Bacon onward. Milton, "Induction before Hume," 61. See also Richard H. Popkin, *The History of Skepticism from Erasmus to Descartes* (Assen: Van Gorcum, 1960).

35   For discussions of this point see Campbell, *Truth and Historicity*, 174; Milton, "Induction before Hume," 70–73. See also Poovey, *A History of the Modern Fact*, 106–10.

36   Campbell, *Truth and Historicity*, 203; Buckle, "British Sceptical Realism," 8, 17.

37   John Locke, *An Essay Concerning Human Understanding* (New York: Prometheus Books, 1995), IV.III.29, p. 456. See also Buckle, "British Sceptical Realism," 8; Woolhouse, *The Empiricists*, 81–84.

38   See, for example, Milton, "Induction before Hume," 69; Buckle, "British Sceptical Realism," 25.

39   Buckle, "British Sceptical Realism," 20.

40   For extensive demonstration of this point see Galen Strawson, *The Secret Connexion: Causation, Realism, and David Hume* (Oxford: Clarendon Press, 1989).

41   Buckle, "British Sceptical Realism," 25.

42   See David Fate Norton, *David Hume: Common-Sense Moralist, Skeptical Metaphysician* (Princeton: Princeton University Press, 1982); David Fate Norton, ed., *The Cambridge Companion to Hume* (Cambridge: Cambridge University Press, 1993).

43   The fact that Hume's moral theory was based on his epistemology and hence on a rejection of any rational sources of action, that he viewed morality as an entirely human affair, and that he argued that our notion of things like "justice" were artificial social conventions arising with private property gave plenty of cause for alarm at the time. See Norton, *David Hume*, 9, 172; Hume, *A Treatise of Human Nature*, Book III.

44   For a useful explication of this fundamental point see Keith Lehrer, "Reid, Hume and Common Sense," *Reid Studies*, 2, no. 1 (Autumn 1998): 15–25. Reid was not the first to respond in this manner to Hume, however. See Henry Home and Lord Kames, *Essays on the Principles of Morality and Natural Religion* (Edinburgh: Kincaid & Donaldson, 1751), esp. Part II, discussed at length in Chapter 7.

45   Thomas Reid, *An Inquiry into the Human Mind on the Principles of Common Sense*, ed. Derek R. Brookes (Edinburgh: Edinburgh University Press, 1997), 4. Henceforth cited as *Inquiry*. Some subsequent page citations to this book will be made in the body of the text.

46   And indeed elsewhere in the *Inquiry* he mentions *only* science and common sense as being threatened by the theory of ideas (ibid., 75–76). Although some scholars argue that in Reid's system God guarantees, in Cartesian fashion, the reliability of our perceptions, others maintain, cogently in my view, that although Reid may be said to ascribe to a form of "providential naturalism," it is the second term, rather than the first, that informs his epistemology. For an instance of the former position, see David Fate Norton, *David Hume: Common-Sense Moralist, Skeptical Metaphysician* (Princeton: Princeton University Press, 1982), 203. For the latter see Derek Brookes, "Introduction," in Reid, *Inquiry*, xxi–xxii; Patrick Rysiew, "Reid and Epistemic Naturalism," *Philosophical Quarterly*, 52, no. 209 (October 2002): 437–56. Lehrer and Warner have argued that rather than supplementing "his naturalistic epistemology supernaturally with the assumption of God's existence [...] Reid advanced a principle of the trustworthiness of our faculties which suffices for justification of our common-sense beliefs without appeal to the existence of God." Keith Lehrer and Bradley Warner, "Reid, God and Epistemology," *American Catholic Philosophical Quarterly*, 74, no. 3 (Summer 2000), from an abstract of the article. My own reading of Reid's works confirms this perspective. Reid's epistemological arguments are clearly intended to stand alone as scientific analyses of the human mind. Although he locates his conclusions in a theistic framework, he does not base them upon it. There can be little doubt, on the other hand, that Reid's religious concerns were important and fundamental to his overall perspective.

47 Newton's *regulae philosophandi* were: (1) No more causes of natural things should be admitted than are both true and sufficient to explain their phenomena. (2) Therefore, the causes assigned to natural effects of the same kind must be, so far as possible, the same. (3) Those qualities of bodies that cannot be intended and remitted (i.e., qualities that cannot be increased and diminished) and that belong to all bodies on which experiments can be made should be taken as qualities of all bodies universally. Steffen Ducheyne, "An Editorial History of Newton's *regulae philosophandi*," *Estudios de Filosofía*, 51 (June 2015): 145–48. For more on Reid's philosophy of science, particularly as it relates to the scientific methodology of Isaac Newton, see L. L. Laudan, "Thomas Reid and the Newtonian Turn of British Methodological Thought," in *The Methodological Heritage of Newton*, ed. R. E. Butts and J. W. Davis (Toronto: University of Toronto Press, 1970), 103–31; Callergård, *An Essay on Thomas Reid's Philosophy of Science*.

48 In the sense of "this is how we are made" or "this is the architecture of our mind/ brain"—which provides the basic structures for mental activity. This is not to say that Reid believed all such hardwired features of our mind are immediately active once we are born and do not mature and develop along with other innate human features; rather, he believed that this is the basic mental equipment that, barring some genetic or developmental abnormality, all human beings possess. Reid believed that our minds are active and in many respects quite malleable and obviously capable of learning all sorts of things, but such learning would not be possible without the basic features of common sense.

49 According to Reid (quoting Hume), Hume denies causality any "'mark of [...] intuitive certainty.'" *Intellectual Powers*, 500.

50 See, for example, *Inquiry*, 195–97; *Intellectual Powers*, 490–93; Thomas Reid, *Essays on the Active Powers of Man* (Edinburgh: Bell & Robinson, 1788), 289–90. Henceforth cited as *Active Powers*.

51 Aberdeen University Library, MS 2131/8/I/1. Or as Reid put it elsewhere, Hume's argument that as far as we know, things may begin to exist without a cause is the most "shocking paradox" that he had ever advanced. *Active Powers*, 32.

52 See, for example, Peter J. Diamond, *Common Sense and Improvement: Thomas Reid as Social Theorist* (Frankfurt am Main: Peter Lang, 1998); Richard Olson, *Scottish Philosophy and British Physics 1750–1880: A Study in the Foundations of the Victorian Scientific Style* (Princeton: Princeton University Press, 1975), 45–50, 107, 182–83.

53 See, for example, Keith Lehrer, *Thomas Reid* (London: Routledge, 1989), 38, 207, 255–60; Dale Tuggy, "Thomas Reid on Causation," *Reid Studies*, 3, no. 2 (Spring 2000): 3–27; Robert Callergård, *An Essay on Thomas Reid's Philosophy of Science* (Stockholm: Stockholm University, 2006), 18–20. My disagreement here and in the note above is more one of emphasis than substance. Readers interested in exploring Reid's philosophy of science in greater detail—and in particular its Newtonian dimensions—should consult Callergård's *Essay*.

54 *Intellectual Powers*, 238; see also 101–3, 503–4; Aberdeen University Library MS 2131/ 6/III/5.

55 *Intellectual Powers*, 422.

56 As discussed above, Reid counts causality as a first principle of necessary truths.

57 *Intellectual Powers*, 497.

58 *Active powers*, 31–32, 46–47, 286–88; *Intellectual Powers*, 101–5, 215–16, 497–500; *Inquiry*, 59, 122, 190.

59   *Inquiry*, 122; *Intellectual Powers*, 51–52; 101–3; 201–2; 215–16.

60   *Intellectual Powers*, 499. See also Baruch Brody, "Hume, Reid, and Kant on Causality," in *Thomas Reid: Critical Interpretations*, ed. Stephen F. Barker and Tom L. Beauchamp (Philadelphia: Philosophical Monographs, 1976), 13.

61   See Hume, *A Treatise of Human Nature*, 271; Reid, *Inquiry*, 41.

62   *Intellectual Powers*, 499. This trope occurs frequently in Reid's writings. See, for example, *Intellectual Powers*, 412, 426, 461, 491; *Active Powers*, 29.

63   *Intellectual Powers*, 39.

64   Ibid., 463–64.

65   Ibid., 465.

66   See Christopher J. Berry, *Social Theory of the Scottish Enlightenment* (Edinburgh: Edinburgh University Press, 1997), 23–28.

67   *Active Powers*, 145, 374; *Intellectual Powers*, 57; Aberdeen University Library MS 2131/7/VII/9: "The blessings of life which we enjoy as members of Society are by far more important and make up a greater part of our happiness than all that we can enjoy as individuals."

68   See *Intellectual Powers*, 68–70; *Active Powers*, 447–50. "This interchange of human minds, by which their thoughts and sentiments are exchanged, and their souls mingle together as it were, is common to the whole species from infancy." *Active Powers*, 451.

69   *Active Powers*, 285; *Intellectual Powers*, 216, 240–41. See also *The Philosophical Orations of Thomas Reid*, ed. D. D. Todd (Carbondale: Southern Illinois University Press, 1989).

70   Dear, *Discipline and Experience*, 23. See also Lorraine Daston, "Baconian Facts, Academic Civility, and the Prehistory of Objectivity," *Annals of Scholarship*, 3, no. 4 (1991), 340–42; Milton, "Induction before Hume," 51–55.

71   Dear, *Discipline and Experience*, 44. See also Daston, "Baconian Facts," 348–49, for similar observations.

72   Daston, "Baconian Facts," 342; Dear, *Discipline and Experience*, 18–20. See also Milton, "Induction before Hume," 53.

73   Quoted in Daston, "Baconian Facts," 341.

74   Dear, *Discipline and Experience*, 30, 42–43. Daston refers to this as a form of "epistemological optimism." Daston, "Baconian Facts," 344. See also Campbell, *Truth and Historicity*, 73; Milton, "Induction before Hume," 53. For Aristotle's discussion of the need to proceed, in "special inquiries" like physics, from accepted axioms, see *Metaphysics*, Book IV, Chap. III, 1005, pp. 735–36.

75   For analysis of the sources of credibility in seventeenth-century English science see Steven Shapin, *A Social History of Truth: Civility and Science in Seventeenth-Century England* (Chicago: University of Chicago Press, 1994). For discussion of the rise of numerical representation as a source of credibility, see Poovey, *A History of the Modern Fact*.

76   As Dear makes clear, much of seventeenth-century science—including Galilean science—remained committed to framing individual event-experiments in terms of common, universal experience, and the English Royal Society was in many ways an exception to this. Nevertheless, the cumulative result was a steady movement away from the conclusions of common-sense experience as a basis for scientific generalization, during the seventeenth century. Discussed in greater detail in Chapter 2.

77   Daston, "Baconian Facts," 356.

78   Daston, "Baconian Facts," passim; Dear, *Discipline and Experience*, Chapter 8; Simon Schaffer and Steven Shapin, *Leviathan and the Air Pump: Hobbes, Boyle, and the Experimental Life* (Princeton: Princeton University Press, 1975). As Schaffer and Shapin make clear,

however, the fact-driven paradigm of the Royal Society was not uncontroversial as a basis for social and scientific order.

79 See Dear, *Discipline and Experience*, Chapters 2–5; Daston, "Baconian Facts."

80 See Dear, *Discipline and Experience*, Chapter 8 and Conclusion. Poovey discusses the latter in terms of "gestural mathematics." Poovey, *A History of the Modern Fact*, 191.

81 Milton, "Induction before Hume," 61; Daston, "Baconian Facts," 344–49; Poovey, *A History of the Modern Fact*, 98–99.

82 Dear, *Discipline and Experience*, 6.

83 Daston, "Baconian Facts," 347–48. See also Dear, *Discipline and Experience*, 6, 13.

84 Daston, *Baconian Facts*," 356.

85 Dear, *Discipline and Experience*, 237–40.

86 Including, for example, a strong empiricism, a focus on inductive inference and skepticism toward unwarranted "hypotheses," and a commitment to mathematical handling of some (but not all) physical phenomena.

87 See Jürgen Habermas, *The Structural Transformation of the Public Sphere: An Inquiry into a Category of Bourgeois Society*, trans. T. Berger and F. Lawrence (Cambridge, MA: MIT Press, 1989). For extensive commentaries and bibliographic discussion of Habermas as his work pertains to this chapter see (among others) Craig Calhoun, ed., *Habermas and the Public Sphere* (Cambridge, MA: MIT Press, 1992); Anthony La Vopa, "Conceiving a Public: Ideas and Society in Eighteenth-Century Europe," *Journal of Modern History*, 64 (1992): 79–116; Benjamin W. Redekop, *Enlightenment and Community: Lessing, Abbt, Herder, and the Quest for a German Public* (Montreal: McGill-Queen's University Press, 2000), Introduction and Chapter 1.

88 See, for example, Sharon Achinstein, *Milton and the Revolutionary Reader* (Princeton: Princeton University Press, 1994); John Brewer, *Party Ideology and Popular Politics at the Accession of George III* (Cambridge: Cambridge University Press, 1976), Part III; Mark Knights, *Politics and Opinion in Crisis, 1678–81* (Cambridge: Cambridge University Press, 1994); Nicholas Rogers, *Whigs and Cities: Popular Politics in the Age of Walpole and Pitt* (Oxford: Oxford University Press, 1989); Sophia Rosenfeld, *Common Sense: A Political History* (Cambridge: Harvard University Press, 2011); Kathleen Wilson, *The Sense of the People: Politics, Culture and Imperialism in England, 1715–1785* (Cambridge: Cambridge University Press, 1995).

89 See, among others, Ian Inkster, "The Public Lecture as an Instrument of Science Education for Adults—The Case of Great Britain c.1750–1850," *Paedogogica Historica: International Journal for the History of Education*, 20 (1980): 80–107; Jan Golinski, *Science as Public Culture: Chemistry and Enlightenment in Britain, 1760–1820* (Cambridge: Cambridge University Press, 1992); John Millburn, "The London Evening Courses of Benjamin Martin and James Ferguson, Eighteenth-Century Lecturers on Experimental Philosophy," *Annals of Science*, 11 (1983): 437–55; John Money, "From Leviathan's Air Pump to Britannia's Voltaic Pile: Science, Public Life and the Forging of Britain, 1660–1820," *Canadian Journal of History*, 28 (December 1993): 521–44; Roy Porter, "Science, Provincial Culture and Public Opinion in Enlightenment England," *British Journal for Eighteenth-Century Studies*, 3, no. 1 (1980): 20–46; Simon Schaffer, "Natural Philosophy and Public Spectacle in the Eighteenth Century," *History of Science*, 21 (1983): 1–43; Larry Stewart, *The Rise of Public Science: Rhetoric, Technology, and Natural Philosophy in Newtonian Britain, 1660–1750* (Cambridge: Cambridge University Press, 1992).

90 Stewart, *The Rise of Public Science*, xvi. See also Thomas Broman, "The Habermasian Public Sphere and 'Science *in* the Enlightenment,'" *History of Science*, 36(1998): 123–49.

91   See in particular Shapin, *A Social History of Truth*.

92   Porter, "Science, Provincial Culture and Public Opinion," 29.

93   See, for example, Marina Benjamin, ed., *Science and Sensibility: Gender and Scientific Enquiry 1780–1945* (Cambridge, MA: Blackwell, 1991); Patricia Phillips, *The Scientific Lady: A Social History of Women's Scientific Interest, 1520–1918* (London: Weidenfeild and Nicolson, 1990); Ann Shteir, *Cultivating Women, Cultivating Science: Flora's Daughters and Botany in England 1760–1860* (Baltimore: Johns Hopkins University Press, 1996).

94   Stewart, *The Rise of Public Science*, Chapter 2.

95   Aberdeen University Library, MS 2131/3/II/9 and 10.

96   Aberdeen University Library, MS 2131/8/I/2.

97   Paul Wood, "Science, the Universities, and the Public Sphere in Eighteenth-Century Scotland," *History of Universities*, 13 (1994): 121. See also Paul Wood, *The Aberdeen Enlightenment: The Arts Curriculum in the Eighteenth Century* (Aberdeen: Aberdeen University Press, 1993).

98   Paul Wood, "Science and the Aberdeen Enlightenment," in *Philosophy and Science in the Scottish Enlightenment*, ed. Peter Jones (Edinburgh: John Donald, 1988), 58. See also Paul Wood, "Science and the Pursuit of Virtue in the Aberdeen Enlightenment," in *Studies in the Philosophy of the Scottish Enlightenment*, ed. M. A. Stewart (Oxford: Oxford University Press, 1990), 127–49.

99   Wood, "Science, the Universities, and the Public Sphere," 122. See also Golinski, *Science as Public Culture*.

100   Reid, *Inquiry*, 35–36.

101   The six founding members of the Club were: Robert Trail, John Gregory, David Skene, George Campbell, John Stewart, and Reid. All were academics except for Trail, a clergyman. Data on the Wise Club is contained in Aberdeen University Library MS 145 and 539, in the David Skene Papers, and in the Thomas Gordon Papers, esp. MS 3107/1, which includes notes on the presentations to the Club. Quote is from point 17, "Rules of the Philosophical Society in Aberdeen," Aberdeen University Library MS 145. Published material on the Wise Club includes Stephen A. Conrad, *Citizenship and Common Sense: The Problem of Authority in the Social Background and Social Philosophy of the Wise Club of Aberdeen* (New York: Garland, 1987); H. Lewis Ulman, *The Minutes of the Aberdeen Philosophical Society 1758–77* (Aberdeen: Aberdeen University Press, 1990).

102   Shapin, *A Social History of Truth*. See also Schaffer and Shapin, *Leviathan and the Air Pump*.

103   The popular reception of James Beattie's *Essay on the Nature and Immutability of Truth*, essentially a gloss on Reid's views with strongly moralistic overtones and little of Reid's characteristic subtlety, assured immediate propagation of a number of Reid's ideas to a wide public. However, despite Beattie's obvious moralistic and religious concerns, he also makes plenty of reference to common sense as a foundation of natural philosophy. See James Beattie, *An Essay on the Nature and Immutability of Truth: In Opposition to Scepticism* (London: A. Kincaid and W. Creech, 1773), 63, 131–32, 139–41, 148, 189–91, 367, 483.

104   For the latter see *Intellectual Powers*, 527–41. Although Reid outlines a number of sources of error, his root contention is that error is not the "natural issue" of our faculties, "any more than disease is of the natural structure of the body. Yet, as we are liable to various diseases of body from accidental causes, external and internal; so we are, from like causes, liable to wrong judgments." *Intellectual Powers*, 527.

105 A. A. C. Cooper, Third Earl of Shaftesbury, "Sensus Communis: An Essay on the Freedom of Wit and Humour," in *Characteristicks of Men, Manners, Opinions, Times*, vol. 1 (5th ed., 1732).

106 Francis Hutcheson, *A System of Moral Philosophy* (London: A. Millar and T. Longman, 1755); Redekop, *Enlightenment and Community*, 88–89.

107 See Knud Haakonssen, ed., *Thomas Reid. Practical Ethics: Being Lectures and Papers on Natural Religion, Self-Government, Natural Jurisprudence, and the Law of Nations* (Princeton: Princeton University Press, 1990), esp. 168, 248–60, 277–87; quote from 287. Reid did not go so far as to assign the origin of moral values like justice to public utility, since he believed that such values were in the nature of things and were intuited by the human mind. See Reid, *Active Powers*, Essay V.

108 See Haakonssen's introduction to *Thomas Reid. Practical Ethics*, 63–70.

109 See Haakonssen, *Thomas Reid. Practical Ethics*, 82–85.

110 Haakonssen, *Thomas Reid. Practical Ethics*, 110, 134, 144, 177–80. See also Reid, *Active Powers*. Reid's account of moral first principles is discussed further in the Epilogue.

111 Hume, *Treatise*, Book III; quotes from 469–70.

112 See Daniel Gordon, "Philosophy, Sociology, and Gender in the Enlightenment Conception of Public Opinion," *French Historical Studies*, 17, no. 4 (1992): 885–88. In his essay "Of Commerce," for example, Hume makes a point of distinguishing between the understandings of the "common man" and "philosophers"—it is only the latter who are capable of "subtle and refined" reasonings about the "general course of things." Since, according to Hume, the public good "depends on the concurrence of a multitude of causes" it is philosophers who are equipped to make proper sense of them. Hume, *Essays*, 255. Hume goes on to show how intuitive and accepted understandings of the relationship between individual luxury and the greatness of the state are mistaken and no longer applicable in the modern, commercial era. A similar discussion occurs in "Of Refinement in the Arts." Hume, *Essays*, 271ff.

113 Norton, *David Hume*, Chapter 3. See also Norton, *The Cambridge Companion to Hume*.

114 Hume, *Essays*, 277.

115 Gordon, "Philosophy, Sociology, and Gender"; Chisick, "David Hume and the Common People." For Hume "popular opinion" was a "contagion" that needed to be managed so as not to be deleterious. Hume, *Essays*, 120.

116 See, for example, Reid's remarks in *Active Powers*, 255.

117 In his opening remarks to his Glasgow students, Reid said "I desire no man to follow my opinions implicitly, or any farther than he sees them well founded [...] This is a Right which belongs to every man come to years of understanding." Aberdeen University Library, MS 2131/4/II/10. Elsewhere he stated that "In all matters belonging to our cognisance, every man must be determined by his own final judgment, otherwise he does not act the part of a rational being. Authority may add weight to one scale; but the man holds the balance, and judges what weight he ought to allow authority." *Intellectual Powers*, 528.

118 Beattie, *An Essay on the Nature and Immutability of Truth*, 38–40.

119 Reid, *Inquiry*, 210.

120 Thomas Reid, "A Brief Account of Aristotle's Logic," in *The Works of Thomas Reid, with an Account of His Life and Writings by Dugald Stewart*, vol. 1 (Charlestown: Samuel Etheridge Junior, 1813), 144–45.

121 Reid, "A Brief Account of Aristotle's Logic," 153.

122   Reid, "A Brief Account of Aristotle's Logic," 162.
123   *Intellectual Powers*, 526.

Epilogue

1   James Oswald, *An Appeal to Common Sense in Behalf of Religion* (Edinburgh: Kincaid & Bell, 1766), 312.
 2   James Beattie, *An Essay on the Nature and Immutability of Truth* (Edinburgh: Kincaid & Creech, 1773), 141.
 3   Roger Robinson, "Introduction," in James Beattie, *An Essay on the Nature and Immutability of Truth*, ed. Roger Robinson (London: Routledge, 1996), xxix–xxx.
 4   Dugald Stewart, *Elements of the Philosophy of the Human Mind vol. 1* (Boston: Wells & Lilly, 1814), 15–53; 61–80; 188–95; 252–73; 403–49.
 5   Ibid., 207–9.
 6   Dugald Stewart, *Elements of the Philosophy of the Human Mind vol. 2* (Boston: Wells & Lilly, 1818), 39–40; 51–57.
 7   James McCosh, *The Scottish Philosophy* (London: Macmillan, 1875), 283; Henry Laurie, *Scottish Philosophy in Its National Development* (Glasgow: Maclehose, 1902), 221.
 8   Thomas Brown, *Lectures on the Philosophy of the Mind (2 vols.)* (Edinburgh: Adam & Charles Black, 1851), vol. 1, 325–29; 457–59; vol. 2: 22–90.
 9   See Thomas Brown, *Inquiry into the Relation of Cause and Effect* (London: Henry G. Bohn, 1835), esp. 244–49; 273–83.
10   George Davie, *The Democratic Intellect* (Edinburgh: Edinburgh University Press, 1961), 261.
11   Laurie, *Scottish Philosophy in Its National Development*, 291.
12   See Davie, *The Democratic Intellect*, 260–80; quote from 271.
13   The remark was made by Ferrier's contemporary De Quincey. See Arthur Thomson, *Ferrier of St. Andrews: An Academic Tragedy* (Edinburgh: Scottish Academic Press, 1985), x.
14   See James Ferrier, *Lectures on Greek Philosophy and Other Philosophical Remains vol. II* (Edinburgh: William Blackwood, 1866), 9–19; 82–89; 407–59. For affinities between Ferrier and Reid see Davie, *The Democratic Intellect*; Gordon Graham, "The Scottish Tradition in Philosophy," *Aberdeen University Review*, 58 (1999): 1–12.
15   John Stuart Mill, *An Examination of Sir William Hamilton's Philosophy* (London: Longmans, Green, 1865).
16   George Davie, *The Scottish Enlightenment and Other Essays* (Edinburgh: Polygon, 1991), 62.
17   J. Schneewind, *Sidgwick's Ethics and Victorian Moral Philosophy* (Oxford: Clarendon Press, 1977), 63.
18   Henry Sidgwick, *Lectures on the Philosophy of Kant* (London: Macmillan, 1905), 418; 415.
19   See Broad's comments in George Moore, *Philosophical Papers* (New York: Collier Books, 1962), 6.
20   See E. D. Klemke, "G.E. Moore," in *The Cambridge Dictionary of Philosophy*, ed. Robert Audi (Cambridge: Cambridge University Press, 1995), 506–8; Geoffrey Warnock, "George Edward Moore," in *The Oxford Companion to Philosophy*, ed. Ted Honderich (Oxford: Oxford University Press, 1995), 585.

21 See, for example, George Moore, *Philosophical Studies* (New York: Humanities Press, 1951), 57–58; 86–89; John Passmore, *A Hundred Years of Philosophy* (London: G. Duckworth, 1957), 212; Ronald Beanblossom, ed., *Thomas Reid's Inquiry and Essays* (Indianapolis: Hackett, 1983), xliii–xlvii.

22 Moore, *Philosophical Papers*, 33.

23 Ibid., 44.

24 Kuehn argues that Reid's thought exerted a significant impact on eighteenth-century German philosophy, yet admits this very point. See Manfred Kuehn, *Scottish Common Sense in Germany, 1768–1800* (Kingston: McGill-Queen's University Press, 1987), 243–44. For a critique of Kuehn's argument see John Wright, "Critical Notice of [...] *Scottish Common Sense in Germany, 1768–1800,*" *Reid Studies,* 2 (1998): 49–55.

25 Quoted in Kuehn, *Scottish Common Sense in Germany,* 173.

26 See Benjamin W. Redekop, *Enlightenment and Community: Lessing, Abbt, Herder, and the Quest for a German Public* (Montreal: McGill-Queen's University Press, 2000).

27 Kuehn, *Scottish Common Sense in Germany,* 52–64.

28 Joseph Priestley, *An Examination of Dr. Reid's Inquiry ... Dr. Beattie's Essay ... and Dr. Oswald's Appeal* (London: J. Johnson, 1774).

29 Kuehn, *Scottish Common Sense in Germany,* 70–75.

30 Johann Feder, *Logik und Metaphysik* (Wien: Trattnern, 1783), 130–58; quote from 117.

31 I could find only two references to Reid in the book. See Feder, *Logik und Metaphysik,* 50, 158.

32 Kuehn, *Scottish Common Sense in Germany,* 76–80; quote from 76.

33 Ibid., 104–13; Johann Eberhard, *Allgemeine Theorie des Denkens und Empfindens* (Berlin: Voß, 1776).

34 Kuehn, *Scottish Common Sense in Germany,* 119.

35 Johann Tetens, *Philosophische Versuche über die menschliche Natur vol. 1* (Leipzig: Weidmanns Erben u. Reich, 1777), 393; 375–76.

36 Kuehn, *Scottish Common Sense in Germany,* 208.

37 Sidgwick, *Lectures on the Philosophy of Kant,* 408–10; Kuehn, *Scottish Common Sense in Germany,* 170.

38 Immanuel Kant, *Prolegomena zu einer jeden künftigen Metaphysik* (Riga: Hartknoch, 1783), 12.

39 They include: *gemeinen Menschenverstand, gesunden Verstand, geraden* or *schlichten Menschenverstand,* and *gemeinen Verstand.* See ibid., 11–12.

40 For more see Redekop, *Enlightenment and Community*; Jonathan Knudsen, "Friedrich Nicolai's 'wirkliche Welt': On Common Sense in the German Enlightenment," in *Mentalitäten und Lebensverhältnisse: Beispiele aus der Sozialgeschichte der Neuzeit* (Göttingen: Vandenhoeck & Ruprecht, 1982), 87–89.

41 George Giovanni, "Hume, Jacobi, and Common Sense," *Kant Studien,* 89 (1998): 44; Friedrich Jacobi, *The Main Philosophical Writings* (Montreal: McGill-Queen's University Press, 1994), 266.

42 Kuehn, *Scottish Common Sense in Germany,* 145–52.

43 G. W. F. Hegel, *Vorlesungen über die Geschichte der Philosophie vol. 4* (Hamburg: Felix Meiner Verlag, 1986), 140–46; 376–77.

44 F. Moore, "Une copie mal déguisée," in *Victor cousin, les idéologues et les Ecossais* (Paris: Presses de l'Ecole Normale Supérieure, 1985), 37–47.

45   Frederick Copleston, *A History of Philosophy vol. IX* (New York: Doubleday, 1994), 19–36; quote from 29.

46   Joseph Gérando, *Histoire comparée des systémes de philosophie vol. IV* (Paris: Librairie philosophique de Ladrange, 1847), 191–209; Emile Boutroux, *Etudes de l'histoire de la philosophie* (Paris: Germer Bailliére et Cie, 1897), 417–20.

47   Paul Royer-Collard, *Les Fragments philosophiques de Royer-Collard* (Paris: Félix Alcan, 1913); quote from 196.

48   Ibid., 181–88.

49   Ibid., 182, 189.

50   Ibid., 188.

51   For Royer-Collard's critique of "modern philosophy" see ibid., 25–40; 81–105; 213–49; Quote from 105.

52   Victor Cousin, *Lectures on the True, the Beautiful, and the Good* (New York: Appleton, 1870), 33.

53   See the author's preface in ibid.

54   Ibid., 347.

55   Ibid., 40.

56   Ibid., 354–55.

57   Ibid., 49, 353–62, quote from 358. See also Victor Cousin, *Philosophie Ecossaise* (Paris: Librairie Nouvelle, 1857), 395–97.

58   Cousin, *Lectures on the True, the Beautiful, and the Good*, 9–10.

59   Patrice Vermeren, *Victor Cousin: Le jeu de la philosophie et de l'état* (Paris: L'Harmattan. 1995), 95.

60   Ibid., 191.

61   Ibid.; Emile Boutroux, *Etudes de l'histoire de la philosophie*, 438–39.

62   Jean-Jacques Goblot, "Jouffroy et Cousin," in *Victor Cousin, Homo Theologico-politicus*, ed. Eric Fauquet (Paris: Editions Kimé), 69–82.

63   See Théodore Jouffroy, "Translator's Forward" and i–lxvii, in *Esquisses de philosophie morale* (Paris: A Johanneau, 1833); Théodore Jouffroy, *Nouveaux mélanges philosophiques* (Paris: Hachette, 1872), 1–155; Théodore Jouffroy, "De la philosophie et du sens commun," in *Philosophie, France, XIXème siècle: écrits et opuscules*, ed. Stéphane Douaillier (Paris: Librairie Générale Française, 1994).

64   Jouffroy, *Nouveaux mélanges philosophiques*, 141–42; 151–55.

65   See Jouffroy, *Esquisses de philosophie morale*, i–ii; lxiv–lxvii; Théodore, Jouffroy, ed. and trans., *Œuvres Complètes de Thomas Reid vol. 1.* (Paris: Victor Masson, 1836); Jouffroy, *Nouveaux mélanges philosophiques*, 74; Jouffroy, "De la philosophie et du sens commun," 72–74.

66   Boutroux, *Etudes de l'histoire de la philosophie*, 440–43.

67   James Manns, *Reid and His French Disciples: Aesthetics and Metaphysics* (Leiden: Brill, 1994); Adolphe Garnier, *Critique de la Philosophie de Thomas Reid* (Paris: Hachette, 1840).

68   August Comte, *Discours sur L'Esprit Positif* (Paris: Librairie Philosophique J. Vrin, 1995), 128.

69   Ibid., 126–48.

70   See Charles Rémusat, *Essais de philosophie* (Paris: Librairie philosophique de Ladrange, 1842), vol. 1: 176–247; vol. 2: 408–48; quote from vol. 2: 589–90.

71   Elizabeth Flower and Murray Murphey, *A History of Philosophy in America* (New York: Capricorn Books, 1977), introduction; Chaps. 4–6.

72   Garry Wills, *Inventing America: Jefferson's Declaration of Independence* (Garden City, NY: Doubleday, 1978), 175–91.

73 Stephen Conrad, "Polite Foundation: Citizenship and Common Sense in James Wilson's Republican Theory," *Supreme Court Review* (1984): 359–88; Henry May, *The Enlightenment in America* (New York: Oxford University Press, 1976), 205–7; 348–49.

74 Douglas Sloan, *The Scottish Enlightenment and the American College Ideal* (New York: Teachers College Press, 1971).

75 Richard Petersen, "Scottish Common Sense in America, 1768–1850: An Evaluation of Its Influence" (PhD Dissertation: American University, 1964), Chap. III; Sloan, *The Scottish Enlightenment and the American College Ideal*, Chap. IV; Thomas, Miller, ed., *The Selected Writings of John Witherspoon* (Carbondale: Southern Illinois University Press, 1990), 152–73; 229–30.

76 Petersen, "Scottish Common Sense in America," 63.

77 Sloan, *The Scottish Enlightenment and the American College Ideal*, Chap. V; Petersen, "Scottish Common Sense in America," Chap. IV.

78 Samuel Smith, *Lectures ... On Moral and Political Philosophy* (Trenton: Daniel Fenton, 1812), vol. I: 10–16; 20; 138–39.

79 Sloan, *The Scottish Enlightenment and the American College Ideal*, 148–49n.7; 182–83.

80 E. Holifield, *The Gentleman Theologians: American Theology in Southern Culture, 1795–1860* (Durham: Duke University Press, 1978); D. R. Come, "The Influence of Princeton on Higher Education in the South Before 1825," *William and Mary Quarterly* (October 1945): 369–87.

81 Archibald Alexander, who occupied the Chair of Didactic and Polemic Theology at the seminary from 1812 to 1851, published *Outlines of Moral Philosophy* in 1852, a distillation of four decades of lectures. Although focusing on religious considerations, Alexander cites Reid at several points and the book exhibits a basic common-sense orientation in the tradition of Oswald. In later years Charles Hodge, longtime professor at Princeton and founder of the *Princeton Review*, published the highly influential *Systematic Theology* (1872–73), a book pervaded by common-sense teachings. For more see Archibald Alexander, *Outlines of Moral Science* (New York: Charles Scribner, 1854); Charles Hodge, *Systematic Theology* (New York: Charles Scribner, 1873); Theodore Bozeman, *Protestants in an Age of Science: The Baconian Ideal and Antebellum Religious Thought* (Chapel Hill: University of North Carolina Press, 1977).

82 James McCosh, *The Intuitions of the Mind Inductively Investigated* (New York: Carter, 1860); J. Hoeveler, *James McCosh and the Scottish Intellectual Tradition* (Princeton: Princeton University Press, 1981).

83 Daniel Howe, *The Unitarian Conscience: Harvard Moral Philosophy, 1805–1861* (Middletown: Wesleyan University Press, 1988), 27.

84 B. Rand, "Philosophical Instruction in Harvard University from 1636 to 1906," *Harvard Graduates Magazine*, 37 (1928): 46; Petersen, "Scottish Common Sense in America," 146.

85 Levi Frisbie, *Inaugural Address* (Cambridge: Hilliard & Metcalf, 1817), 10–11.

86 Levi Hedge, *Elements of Logick* (Boston: Hilliard, Gray, 1833), 170.

87 Cousin was in fact the only thinker cited in the sermon, which was titled "The Philosophy of Man's Spiritual Nature in Regard to the Foundations of Faith." See James Walker, *Reason, Faith, and Duty: Sermons* (Boston: Roberts Brothers, 1877), 60.

88 Howe, *The Unitarian Conscience*.

89 Francis Bowen, *The Principles of Metaphysical and Ethical Science Applied to the Evidences of Religion* (Boston: Hickling, Swan and Brown, 1855), vi.

90   Petersen, "Scottish Common Sense in America," Chap. VIII; Alexander Kern, "The
     Rise of Transcendentalism 1815–1860," in *Transitions in American Literary History*, ed.
     Harry Clark (Durham: Duke University Press, 1953), 247–314; Merrell R. Davis,
     "Emerson's 'Reason' and the Scottish Philosophers," *New England Quarterly*, 17 (June
     1944): 209–28.
91   Quoted in Davis, "Emerson's 'Reason' and the Scottish Philosophers," 226.
92   Petersen, "Scottish Common Sense in America," 176.
93   Ibid., 117.
94   Herbert Schneider, *A History of American Philosophy* (New York: Columbia University
     Press, 1946), 242.
95   See Francis Wayland, *The Elements of Intellectual Philosophy* (New York: Sheldon, 1868).
96   See Francis Wayland, *The Elements of Moral Science* (Boston: Gould and Lincoln,
     1856). In some respects Wayland and other moralists discussed in this chapter were
     extending the line of thinking begun by George Turnbull (Chapter 7).
97   Mark Noll, "Common Sense Traditions and American Evangelical Thought,"
     American Quarterly (1985): 232.
98   Schneider, *A History of American Philosophy*, Chaps. 19–21.
99   See Frederick Beasley, *A Search of Truth in the Science of the Human Mind* (Philadelphia: J.
     Maxwell, 1822), esp. 51–65.
100 Schneider, *A History of American Philosophy*, 241; Jay Fay, *American Psychology Before
     William James* (New Brunswick: Rutgers University Press, 1939), 91.
101 For example, Thomas Upham, *Elements of Mental Philosophy* (New York: Harper,
     1856), vol. 1: 59–134; quote from iii.
102 Schneider, *A History of American Philosophy*, 241.
103 See, for example, Noah Porter, *The Elements of Intellectual Science* (New York: Charles
     Scribner's, 1890), 103; 198; 391–440.
104 See Louis Menand, *The Metaphysical Club* (New York : Farrar, Straus and Giroux,
     2001). Menand neglects to consider the role played by common-sense philosophy in
     the rise of pragmatism.
105 Charles Peirce, *Collected Papers of Charles Sanders Peirce vol. 1* (Cambridge: Harvard
     University Press, 1934), 296–97.
106 Ibid., 297; 347; 354–66.
107 Ibid., 348.
108 Ibid., 295–97; 305; 347; 361–62.
109 Ibid., 355.
110 The present study may be read as supporting Larry Laudan's suggestion that Reid's
     thought was an important conduit for the transmission of Newtonian method, rightly
     understood, into the mainstream of nineteenth-century intellectual culture. See
     L. L. Laudan, "Thomas Reid and the Newtonian Turn of British Methodological
     Thought," in *The Methodological Heritage of Newton*, ed. R. E. Butts and J. W. Davis
     (Toronto: University of Toronto Press, 1970), 103–31.
111 For more on eighteenth-century German sociopolitical conditions, see Redekop,
     Enlightenment and Community.
112 Thomas Reid, *An Inquiry into the Human Mind on the Principles of Common Sense*, ed.
     Derek R. Brookes (Edinburgh: Edinburgh University Press, 1997), 15.
113 Clifford Geertz, "Common Sense as a Cultural System," in *Local Knowledge: Further
     Essays in Interpretive Anthropology* (New York: Basic Books, 1983), 76. Geertz's list
     includes the focus on ordinary language, the development of phenomenology of

everyday life, continental existentialism, and American pragmatism; "all reflect this tendency to look toward the structure of down-to-earth, humdrum, *brave type* thought for clues to the deeper mysteries of existence." Ibid., 77. There has also been growing interest in common sense in the field of psychology—see, for example, Jurg Siegfried, ed., *The Status of Common Sense in Psychology* (Norwood, NJ: Ablex Publishing Corporation, 1994). Suffice it to say that it is beyond the scope of this Epilogue to provide a comprehensive overview of this topic, and readers are encouraged to explore the citations made herein.

114 See Antonio Damasio, *Emotion, Reason, and the Human Brain* (New York: Avon Books, 1994), 94; Daniel Dennett, *Consciousness Explained* (Boston: Little, Brown, 1991); Ray Jackendoff, *Patterns in the Mind: Language and Human Nature* (New York: Harvester Wheatsheaf, 1993), 172. As we have seen, Reid's critique of what he called the "theory of ideas" or the "ideal system" formed a basic conceptual framework throughout his *ouevre*, beginning with his "Philosophical Orations" delivered at graduation ceremonies in King's College, Aberdeen, starting in the early 1750s. See D. D. Todd, ed., *The Philosophical Orations of Thomas Reid. Delivered at Graduation Ceremonies in King's College, Aberdeen, 1753, 1756, 1759, 1762* (Carbondale & Edwardsville: Southern Illinois University Press, 1989), 58–60. For Reid's critique of the model of mind as "mirror" see Reid, *Inquiry*, 91–93; Thomas Reid, *Essays on the Intellectual Powers of Man*, ed. Derek R. Brookes (University Park: The Pennsylvania State University Press, 2002), 31, 93–95.

115 A lively edited collection that engages with some of the more recent research and thinking on this topic is Renée Elio, ed., *Common Sense, Reasoning, and Rationality* (Oxford: Oxford University Press, 2002).

116 Jackendoff, *Patterns in the Mind*, 211; Paul Ekman, *Emotion in the Human Face* (Cambridge: Cambridge University Press, 1982). Reid, *Inquiry*, 59–60, 190–92.

117 Alan M. Leslie and Stephanie Keeble, "Do Six-Month-Old Infants Perceive Causality?" *Cognition*, 25 (1987): 265–88; Alan M. Leslie, "Spatiotemporal Continuity and the Perception of Causality in Infants," *Perception*, 13 (1984): 287–305; Elizabeth Spelke, "Initial Knowledge: Six Suggestions," *Cognition*, 50 (1994): 432–45.

118 Leslie and Keeble, "Do Six-Month-Old Infants Perceive Causality?" 266–70.

119 Ibid., 285.

120 Ibid., 286.

121 Steven Pinker, *How the Mind Works* (New York: W. W. Norton, 1997), 314–15.

122 Steven Pinker, *The Blank Slate: The Modern Denial of Human Nature* (New York: Viking Penguin, 2002).

123 David Premack, "The Infant's Theory of Self-Propelled Objects," *Cognition*, 36 (1990): 1–16; Susan A. Gelman and Gail M. Gottfried, "Children's Causal Explanations of Animate and Inanimate Motion," *Child Development*, 67 (1996): 1970–87.

124 Gelman and Gottfried, "Children's Causal Explanations of Animate and Inanimate Motion," 1971.

125 Premack, "The Infant's Theory of Self-Propelled Objects," 3.

126 See Reid, *Inquiry*, 190; Thomas Reid, *Essays on the Active Powers of Man* (Edinburgh: Bell & Robinson, 1788), 20, 37, 47, 277, 318, 370; Reid, *Intellectual Powers*, 480, 503–6.

127 Paul Bloom, *Descartes' Baby: How the Science of Child Development Explains What Makes Us Human* (New York: Basic Books, 2004), 17.

128 Jackendoff, *Patterns in the Mind*, 178.

129   Susan A. Gelman and Henry A. Wellman, "Insides and Essences: Early Understandings of the Non-Obvious," *Cognition*, 39 (1991): 213–44; Douglas L. Medin and Andrew Ortony, "Psychological Essentialism," in *Similarity and Analogical Reasoning*, ed. Andrew Ortony and Stella Vosniadou (New York: Cambridge University Press, 1989), 179–85.
130   Gelman and Wellman, "Insides and Essences," 239.
131   Ibid., 243.
132   See the discussion in Bloom, *Descartes' Baby*, 46–50.
133   Reid, *Intellectual Powers*, 43–44, 361–62, 495. Reid talks about "subjects" or "substances" rather than "essences," for the most part, but the distinction being made is basically the same.
134   David R. Olson, "Schooling and the transformation of Common Sense," in *Common Sense: The Foundations for Social Science*, ed. Frits van Holthoon and David R. Olson (Lanham: University Press of America, 1987), 327.
135   Lisa M. Oakes and David H. Rakison, "Issues in the Early Development of Concepts and Categories: An Introduction," in *Early Category and Concept Development: Making Sense of the Blooming, Buzzing Confusion*, ed. David H. Rakison and Lisa M. Oakes (Oxford: Oxford University Press, 2003), 19–20.
136   Susan A. Gelman and Melissa A. Koenig, "Theory-Based Categorization in Early Childhood," in *Early Category and Concept Development: Making Sense of the Blooming, Buzzing Confusion*, 330. See also Sandra R. Waxman, "The Dubbing Ceremony Revisited: Object Naming and Categorization in Infancy and Early Childhood," in *Folkbiology*, ed. Douglas L. Medin and Scott Atran (Cambridge: MIT Press, 1999), 233–84.
137   See Medin and Atran, *Folkbiology*.
138   Spelke, "Initial Knowledge," 439. According to Spelke, "Young infants appear to have systematic knowledge in four domains: physics, psychology, number, and geometry" (433). "Initial Knowledge" presents a broad and accessible summary of current research on the features of innate, initial knowledge. For an overview of modern research on the innate mechanisms of visual knowledge see Elizabeth S. Spelke, "Origins of Visual Knowledge," in *Visual Cognition and Action: An Invitation to Cognitive Science vol. 2*, ed. Daniel N. Osherson, Stephen Michael Kosslyn, and John M. Hollerbach (Cambridge, MA: MIT Press, 1990), 99–127.
139   Gelman and Gottfried, "Childrens' Causal Explanations of Animate and Inanimate Motion," 1970; Reid, *Inquiry*, 200.
140   See Mike Oaksford and Nick Chater, "Commonsense Reasoning, Logic, and Human Rationality," in *Common Sense, Reasoning, and Rationality*, 174–214; Gerd Gigerenzer, Jean Czerlinski, and Laura Martignon, "How Good Are Fast and Frugal Heuristics?" in *Common Sense, Reasoning, and Rationality*, 148–73.
141   See Reid, *Inquiry*, 50–51, 191–92; Reid, *Active Powers*, 14–16, 282; Reid, *Intellectual Powers*, 45–46, 466–67.
142   Jackendoff, *Patterns in the Mind*.
143   Ibid., 218.
144   Marc Hauser, *Moral Minds: How Nature Designed our Universal Sense of Right and Wrong* (New York: HarperCollins, 2006).
145   See Bloom, *Descartes' Baby*, 99–154, for an accessible overview of the research on this topic.

146   See Bloom, *Descartes' Baby*; Donald Brown, *Human Universals* (New York: McGraw-Hill, 1991); Frans de Waal, *Primates and Philosophers: How Morality Evolved* (Princeton: Princeton University Press, 2006); Richard Joyce, *The Evolution of Morality* (Cambridge, MA: MIT Press, 2007); Pinker, *How the Mind Works*; Matt Ridley, *The Origins of Virtue: Human Instincts and the Evolution of Cooperation* (New York: Penguin, 1998); Robert Wright, *The Moral Animal: Why We Are the Way We Are* (New York: Random House, 1994).

147   Ernst Fehr and Simon Gächter, "Fairness and Retaliation: The Economics of Reciprocity," *Journal of Economic Perspectives*, 4, no. 3 (Summer 2000): 159–81.

148   Reid, *Active Powers*, 375.

149   Brown, *Human Universals*.

150   Ralph Linton, "Universal Ethical Principles: An Anthropological View," in *Moral Principles of Action: Man's Ethical Imperative*, ed. R. Anshen (New York: Harper, 1952), 645–69.

151   According to Reid, the "natural judgment of conscience in man, that injustice and treachery is a base and unworthy practice" is "a clear and intuitive judgment, resulting from the constitution of human nature." Reid, *Active Powers*, 418. For Reid's discussion "Of the First Principles of Morals," see ibid., 369–75. While arguing for the existence of self-evident principles of morality, Reid was always careful to argue that "our moral conceptions and moral judgments are not born with us," rather they "grow up by degrees, as our reason does." Ibid., 415. Reid thus believed that our moral judgments are intuitive but that they develop in tandem with our rational faculties and experience of the social world.

152   Although it must be said that as we become more aware of the complexities of human nature, it becomes harder to defend a reductive and repressive morality based on its features anyway. See Jackendoff's perceptive comments on this score in *Patterns in the Mind*, 218–20. For a very wide-ranging and comprehensive discussion of the relationship between emerging understandings of human nature and morality, see Wright, *The Moral Animal*, Part Four. For Reid's rejection of the naturalistic fallacy see Reid, *Active Powers*, 125–26; Reid discusses the difference between natural and moral laws in Aberdeen University Library MS 3061/1/1,16.

153   In Reid's view, "When the education which we receive from men, does not give scope to the education of Nature, it is wrong directed; it tends to hurt our faculties of perception, and to enervate both the body and the mind." Reid, *Inquiry*, 202.

154   See the excellent study on this topic by Laurence Viennot, *Reasoning in Physics: The Part of Common Sense* (Dordrecht: Kluwer, 2001).

155   See Frits van Holthoon and David R. Olson, eds., *Common Sense: The Foundations for Social Science* (Lanham: University Press of America, 1987); S. Caroline Purkhardt, *Transforming Social Representations: A Social Psychology of Common Sense and Science* (London: Routledge, 1993). The latter book makes the case for a strong "two-way interaction" between common sense and science, and the social–psychological contours of scientific discovery.

156   Jacob Bronowski, *The Common Sense of Science* (Cambridge, MA: Harvard University Press, 1978), 39, 69, 70, 75–76. A book whose substance and tenor anticipates Bronowski's is Bertrand Russell's *Our Knowledge of the External World*, first published in 1914 and revised in 1928. Russell seeks to give a scientific account of the world revealed by modern physics in terms of sense experience alone, without making

common-sense assumptions, for example, about the existence of permanent, stable, external "things." As such, Russell presents a Humean analysis of modern physics. Russell, like Bronowski, argues that prior to the twentieth century, common sense and physics were aligned and built upon certain assumptions about the existence and stability of objects in space and time. Now the task has become "reconstructing the conception of matter without the *a priori* beliefs which historically gave rise to it" (84). Subsequent works such as those (discussed here) by Oppenheimer and Conant may be seen as seeking to add further depth and nuance to the common sense/ science relationship, now that it is been made problematic by modern physics. For his part, Russell acknowledges that we cannot totally transcend common knowledge and beliefs—there is no "superfine" branch of knowledge that completely transcends the whole of the knowledge of daily life: "the most that can be done is to examine and purify our common knowledge by an internal scrutiny, assuming the canons by which it has been obtained, and applying them with more care and precision." Bertrand Russell, *Our Knowledge of the External World* (New York: Mentor Books, 1960), 84, 56–57.

157   J. Robert Oppenheimer, *Science and the Common Understanding* (New York: Simon and Schuster, 1954), 74–75; 5. Another text in this genre, from the same era, is Hermann Bondi's *Relativity and Common Sense: A New Approach to Einstein* (New York: Dover, [1962] 1980). Bondi's focus is more squarely on trying to explain relativity in everyday language.

158   James B. Conant, *Science and Common* Sense (New Haven: Yale University Press, 1951), 33, 49–50. More recently, in a book which attempts to rebut "creation science," Tim Berra makes arguments very similar to Conant. Berra argues that scientific method is not qualitatively different than "logical, common-sense steps" which we use every day in problem-solving. When we flick a switch and the light doesn't come on, we unconsciously form a hypothesis—first that the switch doesn't work, which leads us to flick it a few times, next that the bulb is bad, which leads us to try another one, and so on. "We use these logical, commonsense steps many times each day without thinking about the process. *Scientists use these steps consciously*; how they proceed is dauntingly called the *scientific method*, but the method is not difficult to understand." Tim M. Berra, *Evolution and the Myth of Creationism: A Basic Guide to the Facts in the Evolution Debate* (Stanford: Stanford University Press, 1990), 1.

159   Nathan Isaacs, *The Foundations of Common Sense: A Psychological Preface to the Problems of Knowledge* (London: Routledge, [1949] 1999), 52–55.

160   John Ziman, *Real Science: What It Is, and What It Means* (Cambridge: Cambridge University Press, 2000), 300–301; 316; 313.

161   Ibid., 314.

162   Reid, *Inquiry*, 174.

163   Rob Kaplan, ed., *Science Says: A Collection of Quotations on the History, Meaning, and Practice of Science* (New York: W.H. Freeman, 2001), 101.

164   John Angus Campbell, "Charles Darwin: Rhetorician of Science," in *The Rhetoric of the Human Sciences: Language and Argument in Scholarship and Public Affairs*, ed. John S. Nelson, Allan Megill, and Donald N. McCloskey (Madison: University of Wisconsin Press, 1987) 69, 71–72; passim.

165   E.g. Tim M. Berra, *Evolution and the Myth of Creationism: A Basic Guide to the Facts in the Evolution Debate* (Stanford: Stanford University Press, 1990; discussed in note 158 above).

166   Herman Schneider and Leo Schneider, *Dictionary of Science for Everyone* (London: Bloomsbury, 1990), 198.

167   Kaplan, *Science Says*, 174. See also G. Polya, *How to Solve It: A New Aspect of Mathematical Method* (Princeton: Princeton University Press, 1973). Polya emphasizes the links between common sense and mathematical forms of reasoning.

168   Philip J. Davis and Reuben Hersh, "Rhetoric and Mathematics," in *The Rhetoric of the Human Sciences*, 64, 68.

169   Reid, *Intellectual Powers*, 465.

170   Ziman, *Real Science*, 296–300.

171   Jurg Siegfried, "Commonsense Language and the Limits of Theory Construction in Psychology," in *The Status of Common Sense in Psychology*, ed. Jurg Siegfried (Norwood: Ablex Publishing Corporation, 1994), 10, 29.

172   Thomas Luckmann, "Some Thoughts on Common Sense and Science," in *Common Sense: The Foundations for Social Science*, 194.

173   Hans Bertens defines postmodernism as "a crisis in representation: a deeply felt loss of faith in our ability to represent the real." Hans Bertens, *The Idea of the Postmodern: A History* (London: Routledge, 1995), 11.

174   From the beginning of his literary career Hume thought of himself as a moralist (in the broad sense of the word at the time) rather than a natural philosopher, and there is no indication that his self-understanding ever changed in this regard. See David Hume, *A Treatise of Human Nature*, ed. L. A. Selby-Bigge (Oxford: Clarendon Press, 1978), 8. And as Michael Barfoot has observed, although Hume's debt to the culture of eighteenth-century science was considerable, "The textual evidence for Hume's so-called 'Newtonianism' has recently been re-examined and found to be both limited and ambiguous. We can go further […] it can be argued that his rather brief and undeveloped views were either commonplace or vicarious, and perhaps even inconsistent." Michael Barfoot, "Hume and the Culture of Science in the Early Eighteenth-Century," in *Studies in the Philosophy of the Scottish Enlightenment*, ed. M. A. Stewart (Oxford: Oxford University Press, 1990), 151–90; quote from 160–61.

175   See Thomas Kuhn, *The Structure of Scientific Revolutions* (Chicago: University of Chicago Press, 1970).

INDEX

Milton Keynes UK
Ingram Content Group UK Ltd.
UKHW040631210924
1771UKWH00029B/115